Book Title: Structural and Electromagnetic Scenarios for Firefighter Locator Tracking Systems

Book Author: Anthony D. Putorti Jr; Francine K. Amon; Kathryn M. Butler; Catherine A. Remley; William F. Young; Christina Spoons;

Book Abstract: One of the most important aspects of effective firefighter response to an emergency event is awareness of the location of the firefighters involved, especially in cases with limited visibility due to darkness, heavy smoke, or unfamiliar and changing environments. Location and tracking systems (LTS) have been developed and are being refined to aid firefighting operations or the rescue of firefighters in distress. In this National Institute of Standards and Technology (NIST) technical note, LTS technologies are examined with the goal of establishing structural and electromagnetic scenarios that are representative of situations in which firefighters are most in need of this technology. Firefighter injury and fatality data are studied to determine the building and occupancy types that are associated with the highest risk of injuries. Current radio frequency (RF) regulations are explored to provide guidance on the electromagnetic landscape in which LTS are expected to operate on the fire ground. The potential effects of RF attenuation, RF multipath, and RF interference, which impact the ability of LTS to operate and communicate with incident command posts, are also discussed. Notional building and electromagnetic scenarios are presented to support the development of future test methods and standards that will appropriately challenge and evaluate LTS performance. These scenarios are also useful for fire departments and local jurisdictions in determining which types of firefighter LTS may be most effective in the types of structures and occupancies in their community.

Citation: NIST TN - 1713

Keywords: Firefighter; fighter; locator; locate; track; tracking; tracker; scenario; interference; attenuation; multipath

Structural and Electromagnetic Scenarios for Firefighter Locator Tracking Systems

Anthony Putorti Jr.
Francine Amon[*]
Kathryn Butler
Fire Research Division
Engineering Laboratory

Kate A. Remley
William F. Young
Electromagnetics Division
Physical Measurement Laboratory

Christina Spoons
Ashford University

National Institute of
Standards and Technology
U.S. Department of Commerce

_____

[*] Current affiliation: SP Technical Research Institute of Sweden, Brandteknik Forskning, Box 857, SE-501 15 Borås, Sweden

NIST Technical Note 1713

Structural and Electromagnetic Scenarios for Firefighter Locator Tracking Systems

Anthony Putorti Jr.
Francine Amon
Kathryn Butler
Fire Research Division
Engineering Laboratory

Kate A. Remley
William F. Young
Electromagnetics Division
Physical Measurement Laboratory

Christina Spoons
Ashford University

September 2011

U.S. Department of Commerce
Rebecca M. Blank, Acting Secretary

National Institute of Standards and Technology
Patrick D. Gallagher, Under Secretary for Standards and Technology and Director

National Institute of Standards and Technology Technical Note 1713
Natl. Inst. Stand. Technol. Tech. Note 1713, 83 Pages (September 2011)
CODEN: NTNOEF

Disclaimer

Certain commercial entities, equipment, or materials may be identified in this document in order to describe an experimental procedure or concept adequately. Such identification is not intended to imply recommendation or endorsement by the National Institute of Standards and Technology, nor is it intended to imply that the entities, materials, or equipment are necessarily the best available for the purpose.

Abstract

One of the most important aspects of effective firefighter response to an emergency event is awareness of the location of the firefighters involved, especially in cases with limited visibility due to darkness, heavy smoke, or unfamiliar and changing environments. Location and tracking systems (LTS[†]) have been developed and are being refined to aid firefighting operations or the rescue of firefighters in distress. In this National Institute of Standards and Technology (NIST) technical note, LTS technologies are examined with the goal of establishing structural and electromagnetic scenarios that are representative of situations in which firefighters are most in need of this technology. Firefighter injury and fatality data are studied to determine the building and occupancy types that are associated with the highest risk of injuries. Current radio frequency (RF) regulations are explored to provide guidance on the electromagnetic landscape in which LTS are expected to operate on the fire ground. The potential effects of RF attenuation, RF multipath, and RF interference, which impact the ability of LTS to operate and communicate with incident command posts, are also discussed. Notional building and electromagnetic scenarios are presented to support the development of future test methods and standards that will appropriately challenge and evaluate LTS performance. These scenarios are also useful for fire departments and local jurisdictions in determining which types of firefighter LTS may be most effective in the types of structures and occupancies in their community.

[†] Throughout this document LTS refers to location and tracking systems in either the singular or plural sense, whichever is appropriate for the context in which it is used.

Table of Contents

1   Introduction

One of the most important aspects of effective response to an emergency event is awareness of the locations of the responders (situational awareness). An emergency event may require response from law enforcement, the fire service, emergency medical technicians, hazardous materials (HAZMAT) experts, and military personnel, all of which may have different location and/or tracking needs. In order to make the best strategic and tactical decisions, it is of particular importance that the Incident Commander (IC[‡]) understands and is aware of the locations of personnel under his or her command. The individual first responders benefit from knowing their own locations as well, especially during operations in darkness, heavy smoke, or when they become disoriented in an unfamiliar or changing environment. Location and tracking systems (LTS)[§] are being developed with the goal of providing this ability to emergency responders. The use of LTS by the fire service, especially for the rescue of distressed firefighters, is the focus of this technical note.

There are two essential components of LTS performance: 1) the accuracy with which the system determines the location of the firefighter or emergency responder, and 2) the ability of the LTS to transmit usable data to an IC post and other responders. These two abilities can be combined to determine the location of a firefighter, to lead a distressed firefighter to safety, or to lead a rescuer to a distressed firefighter. In most cases of interest to the fire service, the LTS would be used at the site of a structure fire and at an IC post positioned in a nearby street or parking lot. Both components of LTS performance can be severely influenced by the operational environment, consisting of the aforementioned structure, IC post, and surrounding area. The virtually infinite number of combinations of geography, building materials, structural configurations, system radio frequency parameters, and sources of electromagnetic interference make attempts to characterize and generalize the operational environment difficult.

In this technical note, building environment and the RF environment are examined with the goal of establishing scenarios that represent situations in which LTS capabilities are needed for firefighter rescue. There are many governmental and non-governmental organizations with interests in LTS technologies from the perspective of reducing firefighter injuries and fatalities. Links to the activities of stakeholder groups and associations are provided and discussed in Section 2. Examples of historical fires where deployment of an effective LTS would likely have reduced the number of firefighter injuries and fatalities are presented in Section 3.

Since it is unlikely that current LTS technologies will be effective in all environments, firefighter injury and fatality data from various sources are analyzed to identify factors associated with disproportionate rates of injuries and fatalities. This information can assist in determining where the use of LTS has the

---

[‡] Throughout this document IC refers to either Incident Commander or Incident Command, whichever is appropriate given the context in which it is used.

[§] Location technologies and tracking technologies both provide current information on the location of the responder. While location technologies may only provide current location, tracking technologies also monitor or track the path used by the responder to arrive at current location. Path information may be used to provide a path to rescue the responder.

greatest potential for reducing firefighter injuries and fatalities. In addition, examination of historical fire data may identify fire incident characteristics that are important to incorporate into the scenarios. Relevant information such as structure type, occupancy classification, construction materials, and cause of death or injury are analyzed and discussed in Section 4. A survey of the current technologies used by LTS, both for location accuracy and signal transmission quality, is provided in Section 5, with the proviso that this is an extremely active area of research and development and some of the details on the systems may become outdated. Tables are provided summarizing the strengths and weaknesses of each technology. The concepts presented in this section will be used for scenario development.

As mentioned above, the ability to communicate data to the IC is an essential component of LTS. RF wireless communications, or simply RF communications, is typically the approach used, at least in part if not the whole, in connecting LTS with the IC. Thus, existing RF standards and regulations that may have a bearing on the development of test methods for LTS performance evaluation are identified and discussed in Section 6. In order to understand the factors affecting the ability to communicate between firefighters in a structure and the IC site, a methodology for classifying RP propagation is presented and discussed in Section 7. Sources of RF interference that may affect the ability of the LTS to communicate to the IC site are discussed in Section 8.

An effective evaluation of LTS performance is based on three general categories of scenarios. The first set, firefighter scenarios, evaluate the ability of LTS to accurately determine the location of the firefighter in distress, and guide a rescuer to the distressed firefighter's location, based on factors such as the construction and configuration of the building, geography, and fire development. The second set of scenarios, the RF propagation scenarios, evaluates the ability of the LTS to communicate with the IC through space and building materials. The third set of scenarios, interference scenarios, evaluates the abilities of the LTS to operate in an environment with RF interference sources such as radio equipment. One method for evaluating an LTS is to choose a firefighter scenario, overlay an RF propagation condition, and overly an RF interference condition. In this manner, a fire department or locality can tailor the specifics of their community (building types and occupancies) with the RF environment present in the locality.

The firefighter LTS scenarios are developed in Section 9. Due to the immaturity of LTS, the firefighter scenarios have not been generalized to provide a technology neutral list of scenarios. Instead, the strengths and weaknesses of existing technologies are identified and used to develop challenging scenarios. The RF scenarios are developed in Section 10. The approach taken here is consistent with model building code and NFPA fire resistance building categorization [1][2]. Due to the more fundamental nature of RF propagation and interference, these scenarios have been developed and organized in a more LTS technology neutral manner as compared to the firefighter scenarios.

The combination of the firefighter scenarios, RF propagation scenarios, and RF interference scenarios support the development of performance evaluation methods based on appropriate challenges to LTS technologies. These scenarios are designed to represent simulated dangerous conditions in which knowledge of a firefighter's location is critical and are intended to guide the development of test methods and performance metrics that define the usefulness of an LTS to the fire service. While the details of the technologies may change, the location and tracking needs of firefighters are anticipated to remain the same, and thus the scenarios described here represent a good starting point for test development even if the LTS technology specifics change in the future.

2   Stakeholders

Although this document is geared towards the structural firefighting community, it has broader implications for multiple audiences and stakeholders. They may have different goals and working environments, but all would benefit from robust, dependable LTS technology. Stakeholders within the firefighting community as well as others outside the community that would potentially benefit from LTS capabilities are described below.

Fire departments, firefighters and firefighter organizations, and non-profits, such as the International Association of Fire Fighters (IAFF), International Association of Fire Chiefs (IAFC), National Volunteer Fire Council (NVFC), National Fallen Firefighters Foundation, and others, strive to improve the working conditions of firefighters and reduce line of duty injuries and deaths. To this end, they are interested in technologies and equipment, such as LTS, that have the potential for reducing firefighter injuries and fatalities. The Report of the National Fire Service Research Agenda Symposium, sponsored by the National Fallen Firefighters Foundation, for example, states that "Location, tracking and the ability to transmit data to and from the firefighter inside buildings are high priority research issues at this time." [3]

Standards development organizations, such as the National Fire Protection Association (NFPA), have an interest in evolving location and tracking technology. The NFPA Technical Committee on Electronic Safety Equipment has had an ongoing task group to examine transmission quality for equipment, including emerging LTS technologies. This committee develops minimum performance standards for any piece of electronic equipment that fire personnel carry into the fire environment.

Worcester Polytechnic Institute (WPI) has hosted an annual workshop on Precision Indoor Personnel Location and Tracking for Emergency Responders for several years [4]. WPI researchers have been working on LTS technology for many years. The annual workshop is an opportunity for them, as well as other researchers, to share their work. The workshop also includes presentations from manufacturers, government agencies, and emergency responders.

Based on their missions, various government agencies have an interest in reducing the potential for firefighters becoming disoriented and lost inside structures.. The National Institute for Occupational Safety and Health (NIOSH), for example, performs investigations of firefighter injuries and fatalities. As a part of these investigations, NIOSH makes recommendations for preventing future injuries and fatalities.

The Department of Homeland Security's (DHS) Geospatial Location Accountability and Navigation System for Emergency Responders (GLANSER) Project is focused on developing and testing an emergency first responder locating system. The goal is for the IC to be able to visually track personnel by using a combination of building diagrams, floor plans, and location technology [5]. The DHS Science and Technology Directorate is also working on the Physiological Health Assessment System for Emergency Responders, or PHASER program. While PHASER does not yet include tracking and location, it does include medical monitoring of first responder personnel. There is potential for the technologies in these two programs to overlap or combine at some point in the future, particularly in the RF communications elements of the systems. DHS also administers the Assistance to Firefighters Grants Program (AFGP) which provides grants to fire departments and funding for research related to fire fighting technology and firefighter safety.

The National Institute of Standards and Technology (NIST) conducts measurement science research to support firefighter technology. Current firefighter technology research areas include electronic equipment, protective clothing ensembles, self-contained breathing apparatus (SCBA), and other personal protective equipment (PPE), as well as firefighter tactics such as the use of positive pressure and natural ventilation.

The United States Fire Administration (USFA) supports the fire service by providing training through the National Fire Academy. The USFA also administers the National Fire Incident Reporting System (NFIRS) which collects fire incident data, including events resulting in firefighter injuries and deaths. The data from NFIRS is used to better understand and quantify the costs of fire in the United States.

Law enforcement and corrections personnel face inherent dangers in their jobs. A tracking and location system would help determine the location and status of personnel. The Communications Technologies Center of Excellence is a National Institute of Justice program focusing on personnel location and tracking. The desired outcome is a device for personnel to wear that can track their location indoors and outdoors, including elevation [6].

Military personnel conduct operations in indoor and outdoor, urban and rural environments, and there is a need to develop positioning capabilities for soldiers operating in the varying environments. The United States Army has been working on a Militarized 3D Locator which is intended to address precision personnel location while operating in urban environments. The Army is also working to develop RF based software to integrate into the radio systems soldiers use in the field.

Manufacturers need to understand the environments and situations in which fire personnel and other emergency responders work in order to develop technology that will suit the needs of the end user. Some considerations to keep in mind include building types, sizes, and materials; the thermal environment (heat, smoke, and water); and operational concerns such as ruggedness, weight, and user interface. RF communications are a critical element of most of these systems, and the manufacturers must provide an RF communication solution that works with their specific LTS solution, and within the intended deployment environments, including in the presence of RF interference.

LTS has many other potential applications, including medical patient tracking and monitoring of skiers in potential avalanche areas.

3 Historic Events

3.1 Introduction

Historic fire incidents that resulted in firefighter fatalities and injuries provide context for the discussions of firefighter LTS. Large events that involve multiple fatalities, like those in Worcester, MA and Charleston, SC, discussed below, gain much attention, but the majority of operational firefighter fatalities do not take place at large loss events. The following past fatal fires include high-profile incidents as well as a sampling of smaller-scale events in which a LTS would have been useful in finding firefighters who became disoriented or lost during firefighting operations. In these cases, LTS could have made the difference between life and death.

3.2 Worcester Cold Storage and Warehouse

Six firefighters lost their lives in a December 3, 1999 fire at the abandoned Worcester Cold Storage and Warehouse building in Worcester, Massachusetts. Firefighters became disoriented inside the building as they were searching for occupants. The structure had been closed in 1989 and all utilities disconnected, but two homeless people were reportedly seen in the building that night. Firefighters sent to find a disoriented crew also became trapped inside the building. It took eight days for fire crews to locate all of the bodies of the six fallen firefighters [7][8].

The original structure, built in 1906, was a six-story building with about 4000 m^2 (43 000 ft^2) of space. An additional six-story structure was built in 1912 against the west wall of the original building. The newer building added 650 m^2 (7000 ft^2) of storage on the third floor and more than 740 m^2 (8000 ft^2) of storage on the fourth through sixth floors. A door had been added to each level to connect the two structures. The two brick buildings were mainly windowless above the second floor, making it more difficult to estimate where crews were inside the building [8].

Reviewing radio transmissions from the event revealed that several crews had trouble communicating their exact position within the building, which made locating them more difficult. Crews also reported activating their PASS devices, yet surviving firefighters reported they did not hear the PASS devices at any time.

3.3 Charleston Sofa Super Store and Warehouse

Nine firefighters lost their lives in a June 18, 2007 fire in a furniture store and warehouse in Charleston, South Carolina. Firefighters became disoriented in the building after fire conditions changed rapidly.

The structure was originally built in the 1960s as a grocery store that had 1590 m^2 (17 100 ft^2) of space. Several expansions were added after the space became a furniture store, including a 650 m^2 (7020 ft^2) addition in 1994 and a 640 m^2 (6940 ft^2) addition in 1995. These two additions were built next to the original building and attached to the exterior walls of the main building. Access openings were added between the original building and each of the additions. A third addition was added in 1996 to the back of the main structure. The third expansion added 1470 m^2 (15 800 ft^2) of warehouse space, and was attached to the main showroom by a 209 m^2 (2250 ft^2) loading dock. The entire facility encompassed more than 4680 m^2 (50 420 ft^2) at the time of the fire [9][10].

A fire began in packing material and discarded furniture outside an enclosed loading dock area. The fire spread to the loading dock, then into both the retail showroom and warehouse spaces. The smoke and combustible gases flowed into the interstitial space below the roof and above the suspended ceiling of the main retail showroom. The extent of fire spread into the interstitial space was not visible to fire fighters in the store. As the smoke layer descended to the floor, crews became disoriented in the rapidly worsening conditions and radioed for assistance, one firefighter activated the emergency button on his radio, and at least one "Mayday" was called. When the front windows were broken out or vented, the inflow of additional air allowed the heat release rate of the fire to intensify rapidly. The fire swept from the rear to the front of the main showroom extremely quickly. Crews sent to rescue the disoriented firefighters were driven back by the fire intensity and rapid fire spread. Rescue crews knew firefighters were trapped inside the structure, but they did not know how many personnel or which individuals were trapped or where [9][10][11].

3.4 Alabama Residential Structure Fire

A volunteer firefighter died in an October 29, 2008 residential structure fire after becoming disoriented. The home was a 200 m² (2100 ft²) single story brick ranch house that was built in 1969. The victim entered alone though a carport and quickly became disoriented in thick smoke. The victim radioed for assistance, but crews were unable to locate him. He was later found in the kitchen approximately 2.7 m (9 ft) from the door through which he had entered. Multiple Self-Contained Breathing Apparatus (SCBA) units were found on the ground outside the home with their integrated PASS devices sounding. This may have contributed to the difficulty in using the PASS alarm signal in order to locate the lost firefighter [12].

3.5 Kansas Residential Structure Fire

A career firefighter died in a May 22, 2010 residential structure fire after becoming separated from his captain in heavy smoke. The victim and a captain entered the 560 m² (6000 ft²) two-story home to search for a resident and a dog. After the crew found the dog and took it to the front door, they continued to search for the resident in worsening conditions (it was later determined the resident was not in the building at the time of the fire). The victim was separated from the captain when he stopped to clear his mask after becoming ill and vomiting. The captain called a Mayday and searched for the victim; two rapid intervention crews (RIC) were also sent to look for the victim, who was found about 11 minutes later and about 7.32 m (24 ft) from where he was last seen [13].

3.6 Texas Commercial Structure Fire

A volunteer fire captain died in a July 3, 2010 fire in an egg processing plant. Eggs were processed, packed, and shipped in the 5400 m² (58 000 ft²) plant. A fire team of two captains (including the victim) tried to access the area they thought to be the seat of the fire by breaching a wall, where they found stacks of wooden pallets blocking their access. They then entered through the front door to find the fire, where they were met with heavy smoke and intense heat and became disoriented. The captains lost contact with their hose line and called a Mayday. They ran out of air while still trying to find the hose line. They then became even more disoriented and separated from one another. One captain was rescued when crews breached a wall and rescued him. The victim was found the following morning [14].

While it is not possible to claim with certainty that an effective LTS would have assisted rescue crews in locating trapped or disoriented fire personnel more quickly and efficiently in the five incidents discussed above, if such a system allows the IC to know the location of each individual on the scene, it could be used to save lives and reduce injuries for both firefighters and civilians.

4 Firefighter Fatality and Injury Data

4.1 Introduction

Firefighter fatality and injury data were examined to provide guidance to the development of firefighter LTS scenarios. By studying the available data regarding firefighters becoming disoriented and lost during structural firefighting operations, it may be possible to determine if there are building types or occupancies that are overrepresented in the loss data. If so, scenarios addressing the overrepresented building or occupancy types may be ranked higher, in terms of relative importance, than other scenarios. In the event that some locator technologies work better than others in different types of structures and occupancies, a fire department or locality may choose one technology over another based on the properties they protect. In addition, the data may be useful to individual fire departments for determining the potential benefits of firefighter LTS based on the structures and occupancies present in their communities. Since the statistics are typically categorized based on fire and building code descriptions of structure types and occupancies [1][2][15], data sorted using these definitions are presented here. This is in contrast with RF classifications of building structures later in the report, where the structures are classified based on the impacts to RF propagation.

4.2 Data Sources

The United States Fire Administration (USFA) compiles firefighter death and injury statistics using a standardized fire department reporting system. The National Fire Incident Reporting System (NFIRS) is the standard reporting system used by fire departments in the United States to report, and keep standardized records of, fires and other incidents. After an incident, fire departments fill out a standard form, which is submitted to a state agency that combines the information from other fire departments in a state database. The state agency transmits the NFIRS data to the National Fire Data Center, where it is stored and available for analysis [16].

The NFIRS program is voluntary, and while approximately 14 000 fire departments participate, not all of the approximately 30 000 fire departments in the U.S. submit data [17]. Therefore, there is the potential for non-uniformity across geographical regions and community sizes, both of which are correlated with fire frequency and severity, leading to potential systematic biases [18]. Despite its limitations, NFIRS is an extremely valuable source of data regarding fire incidents, firefighter injuries, and firefighter fatalities in the United States. During the analysis of firefighter injury and fatality data, researchers typically combine the results of the NFIRS database and data from the NFPA Annual Fire Department Surveys, discussed below, to improve the accuracy estimates of the results.

One method for estimating the potential impact of firefighter LTS on firefighter injuries and fatalities is to examine the overall fatality and injury statistics derived from information received from NFIRS. The fire service casualty module of NFIRS 5.0 contains a field labeled "I_1, Cause of Firefighter Injury." The NFIRS reference guide states that this field should be filled with the code and description for the "immediate cause or condition responsible for the injury" [16]. A list of codes to choose from is also given in the reference guide. From the standpoint of firefighter locators, a related field, "I_2 Factor Contributing to Injury" contains valuable details. In this field, the incident report author is directed to input the code and description for "the most significant factor contributing to the injury of the fire service casualty." One series of codes for this field is "Lost, Caught, Trapped, or Confined" which includes the code for "Lost in Building." The intent of this field is to provide the *most* significant factor contributing to the injury. The reference guide presents a valuable example of an injured firefighter that became disoriented and lost in a building, then suffered a smoke inhalation injury. For this incident, the cause of the injury was exposure to heat and smoke, while the most significant factor contributing to the injury was being lost in the building. The NFIRS data could therefore be searched for the code for "Lost in Building" to estimate the number of fatally or non-fatally injured firefighters that became lost in buildings during firefighting operations. Unfortunately, there is only one entry allowed on the NFIRS form for factor contributing to an injury or fatality, so it is likely that there are incidents where the firefighter(s) became lost, but a different causal factor, such as "roof collapse" or "fire progress," was judged to be more significant. Instances of such coding would tend to underreport instances of lost and disoriented firefighters.

The National Fire Protection Association (NFPA) also compiles firefighter fatality and injury statistics in its Fire Incident Data Organization (FIDO) system [19]. The NFPA compiles firefighter fatality statistics via a census process, which has the goal of capturing every firefighter fatality in the United States. The NFPA becomes aware of firefighter fatalities via a variety of sources, including the USFA, the Public Safety Officers' Benefits Program (PSOB), and news wire services [19]. The NFPA corresponds with the affected departments, and data is compiled from the incident. For each firefighter death, the incident is analyzed and categorized by NFPA staff according to the 1981 edition of NFPA 901 *Uniform Coding for Fire Protection* [20], including a determination of the "cause of injury" and the "nature of injury" [19]. The "cause of injury" is the first of a chain of events leading to the firefighter fatal injury, such as "caught/trapped collapsing roof." For example, if a firefighter is caught and trapped by a collapsing roof, runs out of air, and dies of asphyxiation, the "cause of the injury" is classified as "caught/trapped collapsing roof". The "nature of injury" is classified as "asphyxiation / smoke inhalation." As with the NFIRS system, a weakness of the FIDO system is evident when there are multiple causes, or when it is unknown which cause is the first in the chain. If the firefighter in the previous example received a call to evacuate the structure, but became lost and disoriented prior to the roof collapse, the proper "cause of injury" would be "lost inside building." The analyst, however, may not know that the firefighter was lost prior to the collapse, and misclassify the cause as "caught/trapped collapsing roof". A partial list of causes is shown in Table 1.

One example of multiple causes of firefighter injury is the Charleston, South Carolina Sofa Super Store incident of June 18, 2007 where 9 firefighters were killed while performing firefighting operations at a structure fire (see Section 3.3 of this report). In the NFPA FIDO system, the factor in this case is listed as "caught/trapped collapsing roof," so the fatalities from the incident would not be included in a database search for "lost inside building." Becoming disoriented / lost in the building and rapidly changing fire conditions, however, were also factors in the Charleston fatalities [9][10]. Therefore, it is likely that the number of incidents where a firefighter was lost or disoriented, and who would have benefitted from an LTS, may also be underestimated by a search of the NFPA FIDO data.

In addition to the firefighter fatality census, the NFPA also performs a stratified random sample survey of fire departments each year. Approximately 3000 fire departments are sent a standard survey that requests information on the number of fire incidents, the numbers of deaths and injuries of civilians and firefighters, property (US dollar) losses, community information, and population information [21]. National projections are calculated using the proportion of the total United States population accounted for by the communities of each size that respond to the survey. The NFPA is confident that the actual number of firefighter injuries is within 6.3% of the estimate [22].

NFPA injury data may be combined with NFIRS data to make use of the relative strengths of each data set, *i.e.*, the greater detail of the NFIRS data and the statistical superiority of the NFPA data. An analysis of firefighter injuries has been performed in this manner, and discussed in the written literature [18]. It should be noted that the details of the "Cause of Firefighter Injury" and "Factor Contributing to Injury" have the same limitations as discussed above related to capturing all of the instances where firefighters may have become disoriented or lost. Given this, it is likely that the number of instances where LTS technology would have been beneficial is underestimated by the data. Therefore, while statistics on the number of firefighter injuries in structure fires related to disorientation or becoming lost are not available, a first-order estimate could be derived by examining the general relationship between the number of injuries and the number of fatalities.

Table 1. 2001 to 2010 Firefighter trauma deaths while operating inside structure fires.*

Cause of Firefighter Injury		Fires	% Fires	Fatalities	% Fatalities
Code**	Description				
102	Fell in hole burned in floor	6	7.2	7	6.2
201	Caught/trapped collapsing roof	10	12.0	25	21.7
202	Caught/trapped collapsing wall	2	2.4	3	2.6
203	Caught/trapped collapsing floor	9	10.8	12	10.5
204	Caught/trapped collapsing ceiling	1	1.2	1	0.9
205	Caught/trapped fire progress	24	28.8	32	28.0
207	Caught/trapped flashover	1	1.2	2	1.7
208	Caught/trapped explosion	1	1.2	1	0.9
209	Caught/trapped falling object(s)	1	1.2	1	0.9
210	Caught/trapped between objects	1	1.2	1	0.9
211	Lost inside building	18	21.6	21	18.6
299	Caught/trapped not classified above	2	2.4	3	2.6
320	Struck by falling object(s)	1	1.2	1	0.9
399	Struck by not classified above	1	1.2	1	0.9
411	Smoke/toxic fire products	2	2.4	2	1.8
602	Jumped from wall/ledge/window	2	2.4	3	2.6
Total		82	98.4	116	100

* Source: Fire Analysis and Research Division, National Fire Protection Association, Quincy MA. Does not include events of September 11, 2001.

** Subset of "Cause of Fatal Injury" codes from NFPA 901, 1981 edition.

4.3   Firefighter Fatalities

In response to a request, the NFPA provided brief narratives of structure fires where firefighter(s) were killed by trauma and being lost was the primary event leading to the fatality [23]. In the NFPA 901 system, this would be "Cause of Injury, 211, Lost Inside Building." For the years 2001 to 2010, excluding the events of September 11, 2001, there were 116 firefighter fatalities due to traumatic injuries in structure fires. Of these deaths, 21 were classified as "lost inside building." Using the narratives from the NFPA, the deaths are categorized by building type and occupancy. Classifications by the NFPA are based on NFPA 901-1976 edition, and non-current editions of NFPA 220 [24] and NFPA 251 [25][23]. The results are shown in Table 2 and Table 3.

Table 2. Firefighter deaths while lost in structures: 2001-2010*

Building Construction Type	Firefighter Deaths	% of Firefighter Deaths
Fire Resistive	3	14.3
Unprotected Noncombustible / Limited Combustible	1	4.8
Unprotected Ordinary	5	23.8
Unprotected Wood Frame	12	57.1

* Based on data and summaries from Fire Analysis and Research Division, National Fire Protection Association. Incidents were classified as Code 211 in NFPA 901.

Table 3. Firefighter deaths while lost in structures: 2001-2010

Building Use (Occupancy)*	Firefighter Deaths	% of Firefighter Deaths
Business	4	19.0
Industrial	1	4.8
Mercantile	1	4.8
Residential**	14	66.7
Multi-Use	1	4.8

* Based on summaries from Fire Analysis and Research Division, National Fire Protection Association. Incidents were classified as Code 211 in NFPA 901.

** 12 of the 14 firefighter fatalities occurred in one- and two-family dwellings. Note that (12/21) = 57% of the deaths were in one- and two-family dwellings.

Examination of non-cardiac, trauma-related firefighter deaths during 2001 using publicly available NIOSH investigation data [26] indicates that the NFPA method of classifying the cause of injury does not capture all of the fires where being disoriented and/or lost in a building was a significant causal factor in the death of a firefighter. This is not unexpected, since the goal of the NFPA classification is to determine the initiating event in the chain of events leading to the firefighter fatality. It does help show, however, that the number of deaths where an LTS may be of benefit is underestimated by the above statistics. Examination of the NIOSH reports for 2001, which in itself is only a subset of the total number of fire deaths, indicates that disorientation was a causal factor in eight fatalities, while the NFPA statistical search indicates only one such death for 2001. The NFPA statistics for 2001 to 2010, however, can be used to provide a lower bound of 21 on the number of lost / disoriented firefighter fatalities, while the total number of firefighter fatalities due to traumatic injuries in structure fires, 116, provides an upper bound.

The statistics do tell a consistent story on the types of buildings and occupancies in which firefighters are becoming disoriented and lost. Examination of the NFPA narratives and the data in Table 2 and Table 3, for example, shows that approximately 57% of the firefighter deaths associated with being lost occurred in residential, one- and two-family homes, with building construction classified as unprotected wood frame. For the NIOSH investigations [26] into traumatic firefighter deaths in structure fires in 2001 where disorientation or being lost was a factor, 75% of the fatalities occurred in residential occupancies with building construction classified as unprotected wood frame.

4.4   Firefighter Injuries

While firefighter fatality statistics are important for indicating the potential usefulness of LTS , fireground[**] injuries represent a much larger statistical base for analysis. There are three orders of magnitude more fireground injuries than firefighter fatalities. In addition, while the total number of fires has decreased over time, the rate of injuries per 1000 fires has not changed much over the period of 1981 to 2009 [22]. In the years 2003 to 2006, for example, there were an average of 21 firefighter structural fireground deaths per year [27][28], and an average of 40 270 fireground injuries per year, of which 10 560 average injuries per year were judged to be moderate to severe. Of the injuries judged to be moderate to severe, 1490 average injuries per year were due to burns and/or smoke inhalation [18]. It is possible, therefore, that LTS may help prevent thousands of injuries.

As with firefighter fatalities, the largest percentage of fireground injuries also occurs in one- and two-family residential occupancies, with an average of 20,930 injuries per year, or 61% of the total average fireground injuries from 2003 to 2006. Fireground injuries occurring in residential apartments are second, with an average of 5400 injuries per year, or 16% of the total average fireground injuries from 2003 to 2006 [18]. Table 4 shows that while the largest number of fireground injuries occurs in residential occupancies, the rate of fireground injuries per 100 fires is greatest in industrial, utility, and manufacturing occupancies (11.9 fireground injuries per 100 fires), followed by stores and offices (8.4 fireground injuries per 100 fires), public assembly (7.5 fireground injuries per 100 fires), and one- and two-family dwellings (7.0 fireground injuries per 100 fires). Although the risk of injury is historically greater in some non-residential occupancies, given the overall numbers of firefighter death and injuries, there may be a benefit to deploying LTS technologies that are shown to be effective in residential structures of unprotected wood frame construction. It is important, however, for the users of LTS to be aware of the abilities and limitations of the technology.

[**] The fireground is defined as the fire incident area as well as the area needed by emergency personnel to stage apparatus and mitigate the incident. See NFPA 901, 2006 Ed.

4.5 Future Improvements

In the future, it would be useful to determine the number of incidents where firefighters became disoriented or lost in a structure by examining all of the historical narratives in the NFPA FIDO system. These narratives would include incidents where disorientation may not have been classified as the first in the chain of events leading to the fatal injuries. Additional information on firefighter injuries related to becoming disoriented / lost would improve the assessment of scenarios in which firefighter LTS would be most useful. Improved injury data may be available in the future from the Firefighter Injury Research & Safety Trends (FIRST) project [29], which aims to develop a comprehensive national system for capturing firefighter injuries.

Table 4. Average Annual Fireground Injuries and Injury Rates, 2003 to 2006*.

Occupancy	Fireground Injuries		Fireground Injuries per 100 fires
Assembly	1015	3%	7.5
Educational	280	1%	4.2
Institutional	135	0%	1.9
Residential – 1 & 2 family	20 930	61%	7.0
Residential – Apartments	5400	16%	5.8
Stores & Offices	1925	6%	8.4
Industrial, utility, manufacturing	1375	4%	11.9
Storage	1885	5%	6.1
Special	225	1%	1.0
Total Structures	34 450		6.6

*The data are based on a combination of NFIRS data and the results of the NFPA Annual Fire Experience Survey [18].

4.6 Summary

Based on the information available, the largest number of firefighter deaths associated with becoming disoriented and lost occur at residential structure fires, specifically in one- and two-family dwellings. In addition, these deaths are predominantly associated with structures that are of unprotected wood frame construction. Therefore, the firefighter scenarios should specifically address the performance of LTS in buildings of unprotected wood frame construction and one- and two-family residential structures. In addition, the scenarios should also address occupancies where the risk of injury and death are highest, such as industrial, utility, and manufacturing facilities.

5   Current LTS Technologies

5.1   Introduction

The development of LTS technology has been rapid over the past 10 years and continues to accelerate. Initially, the most common LTS designs were comprised of a primary location technology and a means by which to transmit the data to an IC post (usually via an RF signal). Over the past five years or so it has become apparent that a single location technology is unlikely to perform reliably within the wide range of environments in which the fire service routinely operates. As a result, most LTS under development today are hybrid systems, incorporating two or more location technologies in addition to the data transmission signal. It is therefore difficult to summarize current LTS technologies using a systems-scale approach; in lieu of this approach the individual technologies will be described, along with their strengths and weaknesses. It is assumed that the size, weight, ruggedness, cost, and user interface of each LTS technology will be made acceptable to the emergency response community as the technologies mature. A general discussion of hybrid systems is included after the LTS technologies have been presented.

Before delving into specific LTS technologies, it is useful to clarify a few distinctions regarding LTS functions and operations. First, "location" refers to the position of a person or beacon at a precise moment in time. Tracking refers to the time history of the location of a person or beacon, *i.e.*, knowing the path the person or beacon took to get to their current location. Some LTS do not track people or beacons, offering only their current location. In some cases, a beacon may be placed in a strategic, fixed location to enable the LTS to communicate with other beacons, people or the IC post. Beacons may also be placed on a robot for reconnaissance purposes.

Many LTS technologies benefit from or depend upon the use of floor plans of the structure(s) involved in the emergency event; while floor plans offer a visual aid to the user, they also increase the complexity of the LTS user interface considerably, as well as limiting applicability. It is incumbent upon the fire service to obtain and maintain current floor plans of the structures in their jurisdiction that are most likely to pose a threat to firefighter safety, particularly if their LTS relies upon floor plan accuracy. Maintenance of current floor plans may be integrated into the permitting process for commercial or industrial structures but may prove to be problematic for residential structures due to undocumented owner modifications.

With respect to the LTS user interface, it is assumed that the information presented to the user is acceptable; however, it should be noted that the complexity of the user interface can have a significant impact on the IC's (or firefighter's) ability to interpret the data and react appropriately to it. In some cases, it may be necessary to assign additional personnel to the task of monitoring the LTS at the IC post. As more buildings are constructed with data systems that communicate directly with the fire service, it will become increasingly important that the fire service adapt to meet this challenge.

Construction of new and future structures, particularly commercial and industrial structures, may have fire service LTS and/or communication pathways incorporated into their design. If this is the case, the weaknesses of the technologies described in the following paragraphs may be minimized by design. Even with regulatory support, when considering the lifetime of a commercial or industrial structure, as well as the cost of retroactively installing an LTS/communication system, it will take many years for structurally-integrated systems to become widely established.

Finally, as mentioned above, technologies that are used to locate people or beacons within a structure generally use an RF signal to transmit the location data to an IC post outside the structure; however, there are LTS technologies that use RF signals to establish location within a structure as well. The following technologies are loosely grouped into "non-RF based" if their primary means of determining location is not directly related to RF signals, and "RF-based" if they use some form of RF signal for location purposes. A detailed analysis of RF technology used for communication between a structure's interior and the IC post presumably outside of the structure is the subject of Sections 7 and 8.

There is a wealth of information about LTS technologies, both for location and IC communication purposes, is available on the Worcester Polytechnic Institute's (WPI) website from their annual workshop on <u>Precision Indoor Personnel Location and Tracking for Emergency Responders</u> [4]. This workshop has been taking place at WPI since 2004 and provides perspectives on LTS state of the art from first responders, manufacturers, government agencies, and academia. Unfortunately, there is little information in the literature on this subject. Reasons for this lack of peer reviewed literature largely stem from the pace at which this technology is being developed and the desire of manufacturers to protect their intellectual investment.

5.2 Non-RF-Based Technologies

The technologies described in this section may use RF signals to communicate with an IC post, but they rely primarily upon other techniques to establish the location of a person or beacon. This list covers LTS technologies that have been introduced to the emergency response community via workshops, white papers, journal articles, etc. The list of LTS technologies is constantly growing due to the rate at which new ideas are generated and developed.

5.2.1 Audible Alarms

One of the oldest, least complicated, and well established LTS technologies is the personal alert safety system (PASS). In its simplest form, the PASS device consists of motion detectors and sound generator, and will emit an audible alarm if the user does not move for a specified length of time. Ideally, other firefighters will hear the alarm and use it to locate and rescue the unmoving (downed) firefighter. In addition to the audible alarm, some PASS devices now have RF transmission capability to send an alarm signal to the IC post (one-way communication). Other PASS devices may have two-way RF communication with the IC post, so that the IC can send an evacuation order to the firefighter.

Table 5. Audible Alarms Strengths and Weaknesses

Strength	Weakness
• Simple, robust design • Minimal user involvement	• Alarm sound may not be heard or recognized • Alarm sound may not guide rescuers to downed firefighter • Minimal interaction with IC post

5.2.2 Dead Reckoning

The fundamental principle of dead reckoning is to begin travel from a known position and calculate the current position based on elapsed time and estimates of speed and heading. This principle forms the basis for inertial LTS technology, and may also contribute to other LTS technology development. For fire service use, dead reckoning in its purest form, i.e., estimating paces and headings while carrying out

16

firefighting operations, would be quite difficult due to the stressful, disorienting, low visibility conditions. For this reason dead reckoning is rarely used as a stand-alone location technique by the fire service.

Table 6. Dead Reckoning Strengths and Weaknesses

Strength	Weakness
• Fundamental principle in LTS technology development	• Requires estimates of distance and heading • Errors accumulate over time/distance

5.2.3 Inertial

One of the most prevalent LTS technologies at this time, generally used in combination with other technologies, is the inertial navigation system. Inertial technology uses accelerometers and gyroscopes to estimate the distance traveled and heading of a person. These components, along with data processing modules, are usually packaged into inertial measurement units (IMU), which are worn by the traveler and calculate position from the last known point. As technology evolves, particularly with the advent of micro-machined electromechanical systems (MEMS), the IMU packaging is becoming smaller, lighter, more rugged, more efficient, and less expensive; however, there is still room for improvement in location accuracy calculations. Constant error correction is needed to compensate for drift in the gyroscopes due to environmental noise and the Coriolis effect. If left uncompensated, the drift in position increases with time and rapidly becomes a significant source of error.

Characterization of the movements of firefighters is also essential to inertial technology. The gait (e.g., crawling, walking, running, hauling a hose line, climbing stairs) and the stride lengths associated with these gaits are necessary information for calculation of the position of the person. Also, the effects of such activities as hacking through a wall with an axe or falling through a hole need to be considered.

Most inertial systems use at least one other technology to establish known points whenever possible so that error accumulation is reduced. Global positioning system (GPS), Bluetooth, radio frequency identification (RFID), and RF ranging are all examples of technologies that may be used to provide known points whenever signals are available, although the error correction they provide is only as good as the position obtained from them. If floor plans are available, inertial technology can also be configured to map the floor plan to the motion of responders, which serves as a form of error correction. Progress in development of sophisticated correction algorithms has been rapid in recent years, reducing inertial error accumulation substantially, thus also reducing dependence upon the establishment of known points.

Table 7. Inertial Strengths and Weaknesses

Strength	Weakness
<ul><li>Not vulnerable to signal attenuation</li><li>Not vulnerable to multipath</li></ul>	<ul><li>Accuracy, errors accumulate with time/distance/direction</li><li>Must be calibrated to user's gaits and activities</li><li>Vulnerable to magnetic interference</li><li>Vulnerable to rotational or slew rate errors</li></ul>

5.2.4 Ultrasonic Waves

This technology uses sound waves, at a frequency higher than the human hearing range, to transmit location data from a beacon to a receiver. The receiver is carried by a rapid intervention crew (RIC) member or other rescuer in search of a downed firefighter. The sound waves travel through open air (and smoke), reflecting off solid surfaces (a condition known as multipath), until they make their way to the receiver. The signal strength is used by the receiver to determine the path to the beacon(s), which are carried on the first responder but may also be strategically placed for signal enhancement or for tactical purposes.

Since ultrasonic technology uses open air space for signal propagation, changing conditions within a burning structure may cause the strongest signal to lead the rescue personnel into an area that has collapsed or is otherwise not safe. Conversely, if the downed firefighter can no longer be reached via the path taken to his/her current position, the ultrasonic signal may inform rescue personnel of other available paths. Multipath can have constructive effects, when the ultrasonic waves combine and the signal becomes stronger; or destructive effects, when the waves oppose one another and the signal becomes weaker. It is possible that the structure's geometry can combine with multipath to produce a strong signal that directs the rescue team away from the downed firefighter.

Table 8. Ultrasonic Strengths and Weaknesses

Strength	Weakness
<ul><li>Possible to find the shortest/quickest clear path to the downed firefighter, if one is available</li><li>Beacons can be used for multiple purposes</li></ul>	<ul><li>Conditions may change such that there is no longer a clear air space suitable for signal transmission</li><li>Signal range may be reduced by building configuration</li><li>Multipath may cause the strongest signal to lead the receiver in the wrong direction</li></ul>

5.2.5 Other Supporting Technologies for Non-RF-based Technologies

As mentioned at the beginning of Section 5.2, the above list is not exhaustive. There are some sensor technologies that play a supporting role for LTS and which bear mentioning in the interest of completeness. Barometric measurements provide the third dimension (z-axis) for some systems, magnetic sensors are sometimes used to provide corrections to inertial gyroscopes, and Doppler velocity measurements have been used to improve location accuracy. Map matching, 3-dimensional laser mapping, various optical systems, compasses, and assumptions about straight line paths and 90 ° turns have also been proposed as a means to improve LTS performance.

5.3   RF-Based Technologies

The technologies described in this section use RF signals in some way to establish the location of a person or beacon and/or communicate with an IC post. RF signal characteristics and performance vary widely with wavelength, modulation format, processing algorithms, and antenna/transmitter/receiver design; however, there are some commonalities that are worthy of discussion prior to presentation of the individual RF LTS technologies. We first discuss some of the trade-offs of the use of one type of RF technology over another, and then describe how this technology is used in some of the existing systems. A much more detailed discussion of RF communication is presented in Sections 7, 8, and 10.

RF signals can be transmitted in an open frequency band that does not require a Federal Communications Commission (FCC) license [30]. Many transmission formats have been developed to handle multiple users in these bands, especially for low-data-rate transmissions such as those that are currently used by LTS. However, in some cases, these bands are over-crowded and, consequently, transmissions can be vulnerable to interference from extraneous sources. Emergency response RF signals can also be transmitted in licensed frequency bands that are allocated for public safety. There are several of these licensed bands available. They offer some degree of freedom from interference, but may require specialized, more expensive equipment and are often limited to narrowband transmissions.

In general, longer RF wavelengths (lower frequencies) can penetrate solid material better, but they require larger antennas. For example, a classic, unloaded quarter-wavelength monopole antenna operating at 1 MHz is 75 m long, while a 1 GHz monopole is only 0.075 m. Clearly, the 1 GHz antenna is much easier to integrate into a portable unit. However, as a portable unit is carried throughout a building, signals at the lower frequencies are less impacted by the physical features of the building, which generally results in lower signal attenuation than at the higher frequencies. As well, for the same amount of transmitted power (energy that couples out of the transmit antenna to the receive antenna), the range on the lower frequency systems is generally greater.

In addition to size, another trade-off faced by RF designers is that the amount of data transfer possible is proportional to the amount of bandwidth available. At higher frequencies, wider bandwidths are more common, and so, typically, more data can be transferred between wireless devices (for example, LTS beacons that act as nodes in an ad hoc network).

From the system design viewpoint, cost and flexibility are often major driving factors. LTS development is no different, and for this reason common RF communication approaches, such as WiFi, Bluetooth, and cell phone technology, are often utilized. Most of these technologies were designed to accommodate a large number of users, performing routine communication activities such as voice communications, email, and text messaging. The design intent of these commercial communication technologies is not to provide ubiquitous coverage inside large buildings, and "dead spots" or areas where RF coverage is weak or non-existent are generally acceptable. While the experience of a dropped call or a delayed text message may be an irritant in a routine conversion, during incident response these missed messages or delays may have more significant consequences.

Finally, in contrast to the commonly used technologies mentioned above, ultra-wideband (UWB) technology is also sometimes chosen for use in LTS. UWB technology makes use of a relatively wide frequency band of radio signals. RF signals are considered to be UWB if the bandwidth is greater than 20 % of the center frequency or greater than 500 MHz. UWB technology can carry a large amount of data over a short range and has a low power requirement. (Note that the reason for the low power is to avoid interfering with other systems operating in the same frequency bands.) Typically the signal is pulsed. UWB is less susceptible to multipath due to the probability that at least some of the frequencies in the band will not reflect or scatter off of the building materials or surfaces. Narrowband interference is also less likely, because only a small part of the useable spectrum is affected.

5.3.1   Global Positioning System

Within the past 10 years the satellite-based Global Positioning System (GPS) has become integrated into many communication technologies in the public and private sectors. The position of a receiver is calculated based upon some combination of the distances and/or angles between the receiver and at least three satellites having line-of-sight communication with the receiver (this calculation is also known as triangulation, trilateration, or multilateration). GPS is an excellent location technology for cases where line-of-sight satellite communication is possible, but it fails, even outdoors, when man-made or natural obstructions such as tall buildings or dense forests block the satellite signals. For practical purposes, most indoor location and tracking environments are considered to be "GPS-denied", although GPS signals can sometimes reach an interior receiver through windows or other openings. There are ongoing efforts to bring GPS technology indoors using other transmission technology, such as ultra-wide band signals, and to tightly couple it with other technologies, such as inertial, to make use of complementary strengths.

Table 9. GPS Strengths and Weaknesses

Strength	Weakness
• Excellent performance in unobstructed outdoor environments • Versatile combinations with other technologies are possible	• Requires line of sight communication with at least three satellites • Signal does not penetrate enclosed structures well • Response times can be long • Vulnerable to EMI • Vulnerable to multipath effects • Spatial resolution to 3 meters

5.3.2   Radio-Frequency Identification

RFID technology is comprised of tags containing information about the emergency responder wearing them or the location of the structure to which they are mounted, and a reader that collects the information from the tags and processes it or sends it to a processor. In some cases the tags are stationary and the readers are mobile, and in some cases the reverse is true. Ideally, the tags or readers are permanently mounted throughout a structure prior to an emergency event and the location data is generated whenever the tags and readers are within communication range of each other.

RFID technology has flexibility that makes it attractive for integration with other technologies that can form reconfigurable networks between people and/or beacons (ad hoc mesh networks) or use the tags as beacons (breadcrumbs) to form a line of communication to a signal network or the IC post. Although it is assumed that RFID technology is rugged enough to withstand fireground conditions, some tags are not [31], and deployed tags and/or readers must remain functioning in cases of building collapse.

Table 10. RFID Strengths and Weaknesses

Strength	Weakness
• Flexible interface with other LTS technologies	• RF tags/readers must be able to function in harsh environments • Powerful antennas are needed to read the tags from commonly encountered distances • Breadcrumbs may be destroyed as fire conditions change • Placement of breadcrumbs is subject to cooperation of the firefighter or proper functioning of an automated technique • Vulnerable to EMI

5.3.3   Ranging Radios

Similar to GPS, ranging radios can be used to establish the location of a person or beacon based on the distances and/or angles between them, provided that the location of at least three ranging radios is known and they are able to communicate with one another. Signals used by ranging radios vary widely, but include bandwidth (narrow, wide, ultra-wide), frequency (high, low), and phase (near field, far field); each parameter has its own set of strengths and weaknesses. Depending on the frequency and signal strength of the radios, they may be capable of transmitting through building materials to locate a person or beacon. If the signals cannot reliably penetrate the building or the radios cannot communicate with one another, a trail of repeaters can be used as "breadcrumbs" to retransmit the signal. The ranging radio signal characteristics that work best in one structure type may not be best in another type, therefore it is important for the fire service to understand the limitations of the radio technology with respect to the inventory of structures within their protection zone.

Typically, antennas (infrastructure) are set up around the periphery of the structure using GPS to establish their location; these antennas are then used to determine the location of people or beacons within the structure. Schemes are also being developed that are not dependent on exterior antennas. Deployment of radios, antennas, and repeaters can consume valuable time, although this issue may be mitigated by mounting the external radios to strategically parked emergency response vehicles.

Table 11. Ranging Radio Strengths and Weaknesses

Strength	Weakness
• Versatile choice of signal characteristics	• Infrastructure may need to be deployed upon arrival at emergency event • Building configuration or materials can block signals or reduce effective range • Vulnerable to multipath • Placement of breadcrumbs is subject to cooperation of the firefighter • Vulnerable to EMI • Beacons left along a path may be destroyed as conditions change

5.3.4 Ad Hoc Mesh Networks

Ad hoc mesh networks are communication systems in which firefighters wearing beacons (nodes) communicate with one another, and possibly stationary beacons, to form a flexible network that continually multilaterates the position of each node as the firefighters move throughout the structure. Firefighters are alerted if they move to a location outside the ability of the system to calculate their location. If this condition occurs, a beacon can be placed along the path to ensure that the firefighter's communication is not lost.

As with ranging radios, the signal characteristics of ad hoc mesh networks vary widely from system to system. Also in common with ranging radios, the signal quality strongly depends on the structure type and electromagnetic environment. As the mesh network continually re-forms itself, the data flow must follow new pathways, some of which may be faster or slower than others. Without network redundancy, it is possible that a key network node may lose its signal, causing the nodes that depend on it to also fall out of the network. However, with redundancy, more than one node may be capable of communicating with the IC post, which improves reliability of that very important communication line. Depending on the number of network nodes and the level of redundancy, the amount of data moving through the network can be substantial. However, depending on how the network is configured, the number of nodes, and how often each node checks with other nodes, it is possible that most of the network capacity could be consumed by the nodes maintaining the mesh network and not transmitting data from firefighter to IC post.

Table 12. Ad Hoc Mesh Network Strengths and Weaknesses

Strength	Weakness
• Network can be configured to have inherent redundancy • Possible improved reliability of IC post communication	• Beacons left along a path may be destroyed as conditions change • Placement of beacons is subject to cooperation of the firefighter • One failed node can cause locations of dependent nodes to be lost • Network performance is dependent on management of data flow, can be fast or slow • Vulnerable to EMI • Redundancy reduces network capacity for data flow

5.4 Hybrid LTS Technologies

A common hybrid LTS configuration uses inertial technology as the primary means of locating a person or beacon within a structure combined with an RF signal (GPS, RF, ranging radios) for error correction. The chief differences between these systems lie in the inertial error correction methodology and the degree of integration of the technologies.

More complex systems under development include nearly every technology mentioned in this report, along with virtually every supporting technology. Some systems have multiple error correction options so that at least one of them will provide acceptable accuracy in a given situation. For example, RFID technology has been combined with RF or ranging radios to build beacon networks, either stationary or dynamic, that can lead into areas that RF signals alone cannot penetrate. RFID tags and readers can also be integrated with inertial systems to provide known points for error correction, if they are deployed in a manner that allows their position to be calculated within acceptable accuracy.

5.5 RF Signal Transmission to IC Post

Every LTS must communicate with the IC post and/or the RIC as part of the concept of operations. LTS technology that can provide very precise location/tracking capabilities provides little or no benefit if data cannot reach the IC or RIC. The amount and type of data may vary substantially between different LTS technologies, but in general, RF signals (wireless communication) are the means for data transfer. Thus, the RF signals must be tested under conditions that represent the general characteristics of the channel between the LTS unit and the IC post. System-level considerations for the RF communication link are identical to those faced by the LTS, described in the first few paragraphs of Section 5.3.

No matter what type of system design is used (high or low frequency, narrow or wideband, etc.) the challenges that arise in communicating via RF signals with the IC post generally fall into the categories of RF propagation effects in buildings and RF interference from electromagnetic sources both internal and external to the building. Concrete, steel, metals, glass, stone, brick and other building materials all attenuate the RF signal more so than free space. Signals propagating from basements and below ground structures experience even greater attenuation due to RF losses in the earth. In addition, mechanical features in structures such as the duct work, piping, and metallic finishes along with the steel and structural components can be significant RF reflectors and scatterers that can create a complex multipath propagation environment. The combination of significant attenuation (also called building path loss or simply path loss) and multipath effects can weaken or disrupt the RF communication signal, and prevent communication between LTS devices in the building and the IC post.

RF interference is a potential problem for all RF-based systems, and the IC post presents some unique challenges. An IC post typically includes multiple radio systems, with individual devices transmitting from 1 W to 5 W of power. In certain dense urban environments, repeaters or IC base stations may be used that transmit up to 40 W [32] of power in order to overcome building path loss effects. Unfortunately, these relatively high-power devices can overpower other receive equipment located in the same physical proximity. The use of multiple RF systems, such as LTS, voice communications, RF-based PASS, will typically require different antennas. Significant RF energy can couple between systems if the antennas are not well isolated in an RF sense. For example, if an omnidirectional antenna for a LTS technology that uses a 100 mW, 900 MHz communication system is inadvertently located in the main beam of a 10 dB gain antenna transmitting 1 W of power at 905 MHz, the coupling though the antennas may overload the LTS receiver.

5.6 Summary

Understanding current LTS technology, particularly the strengths and weaknesses of each, is vital to the development of more accurate and reliable systems. From this knowledge, combined with information about the operating environments and situations that trigger the need for rescue, scenarios can be derived that address potential LTS performance problem areas *and* are based on realistic fire service challenges. Because RF technology, used for LTS or for a communications link, plays a key role in most LTS technologies, the following sections of this document focus on possible impairments to the RF signal and how they may be characterized for testing purposes. RF signal path loss and multipath effects measured in large buildings are discussed, along with potential sources of RF interference. Proposed testing scenarios cover these key evaluation criteria of path loss, multipath, and RF interference. Incorporating the information discussed above, as well as insight and testing based on the scenarios that follow below, LTS users can make informed decisions about which systems are more likely to perform well given the particular mix of structures and electromagnetic environments within their protection zone.

6.0 Radio-Frequency (RF) Regulations and Standards

6.1 Introduction

Because most LTS rely on RF technology for communication and/or as part of the LTS itself, this section introduces the regulations and standards that manufacturers of LTS technology must follow. These regulations determine, to a great extent, the choices that manufacturers make in system design. They also help to determine appropriate test scenarios because they dictate, to a great extent, the expected RF environment.

The RF range of the electromagnetic spectrum is a limited resource that is regulated nationally and internationally in order to promote efficient use. Developers, manufacturers, and users of any equipment that emits RF energy must meet the requirements set out by the governing bodies. Today's telecommunications, information technology, and power generation industries are built on technical standards that have been written within the context of the managed RF spectrum. These standards include specifications, guides, practices, and test methods, which support the development and operation of a wide variety of equipment and devices. The combination of RF sources in a locality results in the RF environment in which a fire service LTS must operate, and the LTS itself must follow these rules and standards.

Regulations and standards that govern the use of the RF spectrum, in particular frequency allocation, signal modulation, and limitations on source power and field strength, provide insight into the RF environment that the LTS may encounter. Regulatory limits depend on safety concerns as well as on operational and spectrum sharing considerations. Standards that address electromagnetic compatibility (EMC), that is, the ability of a system to function satisfactorily in an electromagnetic environment containing other sources of electromagnetic interference (EMI), are also important.

This section describes the roles played by regulatory bodies, standards organizations, and professional and nonprofit organizations that are invested in the effective use of the RF spectrum. The purpose of the discussion is to provide for informed development of scenarios for LTS and to support the development of RF tests within the context of LTS technology. The list is intended to be as complete as possible, covering key organizations involved in the RF arena. These organizations interact closely to maintain compatibility. Some relevant online resources are identified.

6.2 RF Spectrum Regulation

Although every country is responsible for management of the RF spectrum within its own borders, global coordination is necessary to harmonize international usage, support interoperable radio communications technology, and address key global issues such as satellite use. In the US, allocation and use of radio frequency bands are regulated by the Federal Communications Commission (FCC) for non-federal (including commercial and public safety) use and by the National Telecommunications and Information Administration (NTIA) for federal government use [33]. Spectrum use is based on the International Table of Frequency Allocations administered by the International Telecommunication Union (ITU). The list below briefly describes the roles of these three organizations.

- FCC: The Federal Communications Commission is an independent United States government agency that regulates interstate and international communications by radio, television, wire, satellite, and cable. Within the FCC, the Wireless Telecommunications Bureau (WTB) manages FCC domestic wireless telecommunications programs and policies, including cellular telephone, paging, personal communications services, and other commercial and private radio services. The Public Safety & Homeland Security Bureau (PSHSB) manages FCC policies pertaining to public-safety communications issues. Rules and regulations to manage the RF spectrum in the U.S. are codified in Title 47 of the Code of Federal Regulations (CFR) [34]. Sections of this code that define sources that could interfere with the operation of fire service LTS include the following:
    - 47 CFR Part 2 Radio frequency allocations and general rules and regulations regarding their use
    - 47 CFR Part 5 Experimental uses, including schools and research
    - 47 CFR Part 15 Radio frequency devices (intentional, unintentional, or incidental radiators) that may be operated without an individual license, including Wi-Fi, Bluetooth, and other wireless systems, RFID, cordless phones, alarm systems, door openers, remote switches, and unlicensed radiators for industrial, scientific, and medical (ISM) equipment [35].
    - 47 CFR Part 18 ISM equipment not intended for radio communication, e.g. microwave ovens, induction cooking ranges, industrial heaters, and medical diathermy equipment
    - 47 CFR Part 20 Commercial mobile radio services
    - 47 CFR Part 22 Public mobile services, including cellular services
    - 47 CFR Part 24 Personal communications services
    - 47 CFR Part 25 Satellite communications
    - 47 CFR Part 27 Miscellaneous wireless communications services
    - 47 CFR Part 73 Radio broadcast services (AM , FM, and TV)
    - 47 CFR Part 76 Multichannel video and cable television service
    - 47 CFR Part 78 Cable television relay service
    - 47 CFR Part 87 Aviation services
    - 47 CFR Part 90 Private land mobile radio services for public safety, industrial/business, and radiolocation applications
    - 47 CFR Part 95 Personal radio services, including wireless medical telemetry
    - 47 CFR Part 97 Amateur radio service
    - 47 CFR Part 101 Fixed microwave services
- NTIA. The National Telecommunications and Information Administration of the U.S. Department of Commerce (DOC) advises the President on telecommunications and information policy issues and regulates the Federal uses of the frequency spectrum. A summary of Federal spectrum use above 30 MHz can be found online [36], and NTIA regulations are published in the Manual of Regulations and Procedures for Federal Radio Frequency Management [37]. The NTIA is developing a handbook on best practices in RF spectrum management, which will include studies on EMI susceptibility and protection [38,39].

- ITU-R. The International Telecommunication Union is the specialized agency of the United Nations that coordinates and standardizes information and communication technologies globally. The Radiocommunication Sector (ITU-R) is responsible for the allocation of frequencies in the global radio spectrum and for coordination of satellite orbits. ITU-R decisions, including the worldwide allocation of frequency bands to radio services and the mandatory technical parameters to be observed by radio transmitters and other RF sources, are published in the Radio Regulations. Frequencies for the use of ISM bands, for example, are designated in RR Nos. 5.138, 5.150, 5.280 and subject to the provisions of RR No. 15.13. Studies of EMI issues are presented in technical reports on spectrum management [40,41].

6.3   Human Exposure Limitations

Limitations on power and field strength from RF sources take into account not only operational considerations but also the health effects of exposure to non-ionizing radiation. In specific cases, the FCC may require a much smaller limit on the allowable power for large transmitters, such as commercial broadcast antennas and cellular base stations, than the maximum value stated in the regulations, due to concerns about exposure. In addition to transmitter power and frequency, human exposure to RF radiation depends on the antenna design, as well as its location and orientation relative to people that may be nearby or working on the equipment.

- FCC regulations include limits for Maximum Permissible Exposure (MPE), based on recommended exposure guidelines from both the National Council on Radiation Protection and Measurements (NCRP) and the American National Standards Institute (ANSI) / Institute of Electrical and Electronics Engineers (IEEE) [42]. Measurement of field strength is required under specified conditions [43]. Typical RF levels have been found to be much lower than the MPE near the base of cellular or PCS cell site towers. It is possible for RF emissions to exceed guideline levels for roof-mounted cellular and PCS antennas, but only on the rooftop itself at close distances to and directly in front of the antenna [44].
- FDA. The Center for Devices and Radiological Health (CDRH) within the U.S. Food and Drug Administration (FDA) is responsible for regulating medical devices and radiation-emitting electronic products. FDA collaborates with the FCC to regulate wireless technology devices such as wireless computer networks and cellular phones, and has the authority to act in cases of hazardous RF emissions. EMC issues in an environment containing multiple RF emitters are discussed in a draft guidance document on the safe and effective use of RF wireless technology in medical devices [45].
- NCRP. The National Council on Radiation Protection and Measurements is a non-profit organization chartered by the US Congress to develop information and recommendations concerning radiation protection. The FCC has adopted the NCRP's recommended Maximum Permissible Exposure limits for transmitter field strength and power density [46].
- EPA. The U.S. Environmental Protection Agency (EPA) sets protective standards limiting radiation exposure. Recommendations from EPA are used by the FCC in developing radiation protection regulations.

6.4   RF Technology Standards

Several organizations develop standards for radio technology, networks, electrical and electronic equipment, and other applications using the RF spectrum. The standards organizations work together closely, using joint committees, shared standards, and other methods to ensure that standards conform with and complement each other.

- ITU-T. The Telecommunication Standardization Sector of the ITU develops technical standards for interconnection of networks and technologies. Protection against interference is the topic of the ITU-T K series of Recommendations.

- IEC. The International Electrotechnical Commission develops consensus-based international standards for electrical, electronic, and related technologies, including the protection of electrical and electronic systems from RF interference. In particular, the IEC has taken a leading role in developing standards and test methods for EMC, including RF emissions and immunity to interference from other RF sources in the environment. The International Special Committee on Radio Interference (CISPR) develops EMC standards to protect radio transmission and reception equipment, domestic appliances, and computers, while Technical Committee 77 (TC77) works on general EMC issues, including protection of networks and power transmission equipment. TC77 is responsible for the IEC 61000 standard on electromagnetic compatibility. The Advisory Committee on Electromagnetic Compatibility (ACEC) guides and coordinates work on EMC throughout IEC.

- ISO. The International Organization for Standardization also works on EMC issues. A large part of the work on electromagnetic sources is carried out by joint committees with IEC or TTU. The TC 021 Committee is responsible for equipment for fire protection and firefighting, including a working group on components using radio transmission.

- ANSI. The American National Standards Institute facilitates and coordinates voluntary consensus standards in the U.S. and promotes the use of U.S. standards internationally. ANSI serves as the U.S. representative in the ISO and (via the U.S. National Committee) in the IEC.

- IEEE. The Institute of Electrical and Electronics Engineers is a technical and professional engineering organization that develops international standards for telecommunications, information technology and power generation products and services. IEEE technical standards on wireless technology have supported the development of Wi-Fi wireless local area networks (WLAN) (IEEE 802.11a/b/g/n), Bluetooth and ZigBee wireless personal area networks (WPAN) (IEEE 802.15.1 and 802.15.4, respectively), and WiMAX broadband wireless access (IEEE 802.16). Exposure limits for RF emissions developed by IEEE [47] were adopted by ANSI and incorporated into FCC regulations.

- ASTM. ASTM International coordinates development of international voluntary consensus standards to support industry and governments. The Subcommittee on Homeland Security Standards (E54.08) is currently working on a standard for RF wireless control of robots for urban search and rescue applications [48].

- NFPA. The National Fire Protection Association develops standards for protection of first responders. Standards that address electronic devices used by firefighters include PASS devices and thermal imaging cameras. The Electronic Safety Equipment Committee is responsible for standards on the design, performance, testing, and certification of electronic safety equipment used by fire and emergency services during emergency incident operations, as well as their selection, care, and maintenance.

- MIL-STD. The U.S. Department of Defense (USDOD) develops standards to achieve interoperability, commonality, reliability, compatibility with logistics systems, and assurance that a device will meet certain requirements [49]. These standards may be adopted by non-defense government organizations, technical organizations, and industry for use in non-military environments. US Navy Signal-to-Noise Program (SNEP) Teams have examined radio-noise and radio-interference problems at more than 40 sites around the globe [50].

6.5   Other Professional and Nonprofit Organizations

- ARRL. The American Radio Relay League is the national association for amateur radio enthusiasts. It supports the amateur ratio community in the U.S. by providing technical advice and assistance, representing its interests before federal regulatory bodies, supporting educational programs, and sponsoring emergency communications services. The ARRL website includes extensive information on the impact of various devices and equipment on amateur radio operation.

- AAMI. The Association for the Advancement of Medical Instrumentation is a nonprofit organization whose mission is to increase the understanding and beneficial use of medical instrumentation through effective standards, educational programs, and publications. AAMI works with FDA to create medical device standards. A guidance document on evaluating the electromagnetic environment in healthcare facilities and minimizing EMI of medical devices in healthcare facilities provides a useful model for similar efforts in other settings [51].

- PSST. The Public Safety Spectrum Trust holds the FCC license for the 700 MHz public safety nationwide broadband spectrum. Its mission is to provide the organizational structure for national public safety leadership to guide construction and operation of an interoperable nationwide public-safety-grade wireless broadband network.

6.6   Resources

Many useful databases on spectrum allocation, FCC licenses, and RF equipment are available on the internet. Described here are a few resources that can provide additional insight in assessing the potential RF interference with LTS.

- FCC Tools [52] include Spectrum Dashboard [53], which displays a browseable chart of the RF spectrum bands and an interactive map for finding information on licensees in a particular county or state, and FM Model for Windows [54], a tool for predicting ground level power density due to FM antenna systems.

- RadioReference.com [55] provides data on radio communications, including a complete frequency database, trunked radio system information, FCC license data by geographic location and frequency, and discussion forums on scanners, amateur radio, equipment, local radio signals, and other RF communication topics.

- Radioing.com/eEngineer [56] provides lists of EMC standards, standards organizations, software, and associations involved with EMC, as well as some basic RF calculators.

As will be shown in subsequent sections, the references above guide manufacturers on design of their systems, helping to limit the number of required scenarios for adequate testing of LTS technology. These regulations and standards also determine, in large part, the expected types and level of RF interference to a specific LTS technology. Again, this can be useful in development of test methods.

7 Radio Frequency (RF) Propagation in Firefighter Environments

7.1 Introduction

When emergency responders enter a large structure such as an apartment or office building, sports stadium, store, mall, hotel, convention center, or warehouse, communication to individuals on the outside is often impaired. The received signal strength from wireless devices, such as radio-linked LTS, is reduced due to path loss (attenuation) caused by propagation through the building materials and scattering by the structural components [57][58][59][60][61][62][63][64][65][66][67][68][69][70]. Also, the large amount of signal variability throughout the structure can present difficulties for the equalizers in receivers in such RF-based communication systems [67].

In order to develop test scenarios that can be used to assess the RF performance of LTS, it is necessary to first understand how key radio-channel impairments such as attenuation, self interference caused by reflections (multipath), and RF interference affect these systems' performance in the field. Typically, this understanding is acquired through measurements made in representative environments, where a sufficient amount of data is collected to form a statistical analysis of the effect of a given impairment. Once the data have been collected and analyzed, scenarios that represent the key impairments may be developed. In this section, we describe a set of measurements made by NIST researchers in which RF-channel data were acquired in representative responder environments. From these data, the key channel impairments of attenuation and multipath are quantified. Finally, broad-brush categories of building structures are defined in terms of RF-channel impairments, as opposed to the traditional classification of structures in terms of construction materials used by, for example, the NFPA [24].

Building penetration of RF signals, where signals are transmitted from outside to within a structure or vice versa, is of primary concern to users of LTS, because signals from a portable unit within a structure must typically be communicated to the IC post outside the structure. As background, refs. [71][72][73][74][75][76][77][78][79][80] provide measurement results of building penetration in a variety of structures. Ref. [81] outlines a measurement campaign at 2.4 GHz and 5.2 GHz carried out in support of a system design effort, while [79] and [80] focus on measurements for public safety. Ref. [82] provides a review of radio propagation measurements up to the year 1990 within public-safety frequency bands, and also proposes a building classification scheme based on RF characteristics. The classification scheme of [82] is more detailed than is desirable for use in scenario-based testing for LTS assessment.

Most of the references cited above were focused on the development of commercial wireless systems. To extend the limited set of work that has been devoted to the specific needs of the public-safety community, we summarize the results of a set of measurements conducted by the Electromagnetics Division of the National Institute of Standards and Technology (NIST). Researchers in the Electromagnetics Division have been involved in a multi-year project to support the development of performance metrics and test methods for RF-based electronic safety equipment used by the public-safety community, such as RF-linked LTS. Part of this work has involved field tests in which measurements were made of radio-propagation-environment characteristics in representative emergency responder environments.

In the results reported here, tests of emergency responder wireless devices were conducted in approximately the same locations where channel characterization tests were carried out. The purpose of these side-by-side tests was to assess the performance of actual wireless devices in the same environment where channel impairments have been measured and assessed. Identifying which radio-channel impairments have the most significant effect on wireless device performance allows development of standardized lab-based test methods that simulate the conditions under which electronic safety equipment will be used in the field. These performance test methods can then be incorporated into consensus standards for such equipment.

In the channel-impairment studies, NIST engineers measured attenuation and multipath, given by the root mean square (RMS) delay spread, in large public structures and other environments where it is expected that wireless communications could potentially be difficult for emergency responders. The characteristics of buildings expected to make wireless communication difficult include those having multiple stories; those with subterranean floors; those with large, deep interior spaces; those with few windows; and outdoor "urban canyons," consisting of city streets surrounded by tall buildings.

From the NIST-measured data, a set of broad classes of building types have been proposed. This classification is based on how well an RF signal can be expected to penetrate and propagate within a given type of structure, as opposed to the structural fire resistance classifications used by the NFPA [24][25] and model building codes[1]. NIST researchers proposed three broad classifications of low, medium, and high attenuation, as well as a classification of "high multipath" for highly reflective environments such as factories.

The NIST Public-Safety Communications Research Lab funded the measurements of the radio-propagation channel impairments. These measurements are summarized in NIST Technical Notes 1540-1542 [61-63], 1545-1546 [64, 65], 1552 [66], and 1557 [69]. The development of building classifications and test methods has been funded by the U.S. Department of Homeland Security's Office of Standards.

We first briefly summarize the locations and measured data that were collected during the NIST field measurements. Details of the measurements and the complete set of data can be found in ref. [70]. Next, we discuss the proposed classification of buildings in terms of RF attenuation and multipath. RF interference, which can also impair an RF-linked LTS transmission, is discussed in Section 8.

7.2   NIST Measurements

To support standards development for wireless in public-safety applications, NIST experiments are focused on the penetration of radio signals from outside to inside a given structure. By reciprocity, signals transmitted from inside to outside are expected to experience the same channel conditions. To simulate an IC station in the channel-impairment studies, a transmit antenna was located outside of each structure at a location where an IC fire truck might be stationed. The receive antenna was placed at various discrete locations within the buildings that could potentially be problematic for radio reception. In each location, the path loss and RMS delay spread (a measure of multipath effects) were measured using a vector-network-analyzer (VNA)-based measurement method described in ref. [70].

NIST tests of emergency safety wireless equipment has focused on RF-based personal alert safety systems (PASS), used by firefighters to indicate when a firefighter is motionless or in distress. We tested RF-based PASS systems from two manufacturers. One operated on a licensed frequency in the 450 MHz band which could be considered a narrowband transmission. The other was a frequency-hopping system [83,84] that operated in the unlicensed frequency band between 902 MHz and 928 MHz. Because many RF-linked LTS are expected to operate in these frequency bands and with similar transmission formats, the conclusions and proposed test methods resulting from the study can be applied to other RF-based emergency safety equipment as well.

7.3   Attenuation and Multipath

In a wireless propagation environment, the amount of attenuation between a base station and user-worn location/tracking device will be affected by many parameters, including the distance between them, the building materials, the location within a given building, the frequency of operation, and the type and orientation of the transmit and receive antennas, among other factors. The large range of possible values for these factors complicate the choice of a representative value for attenuation for classification of structures, the choice and number of realistic scenarios, and the development of lab-based tests.

For the environments we studied, it is clear from the data presented below and described in more detail in [70] that attenuation (path loss) is the dominant failure mechanism for RF transmission. An example of the data is shown in Figure 1, where we have plotted the measured route mean square (RMS) delay spread vs. path loss for a high-rise office building. The success or failure of the RF-based PASS transmission is shown as well, with blue circles representing a successfully received alarm, and red stars representing a failure, defined as a delay of more than one minute.

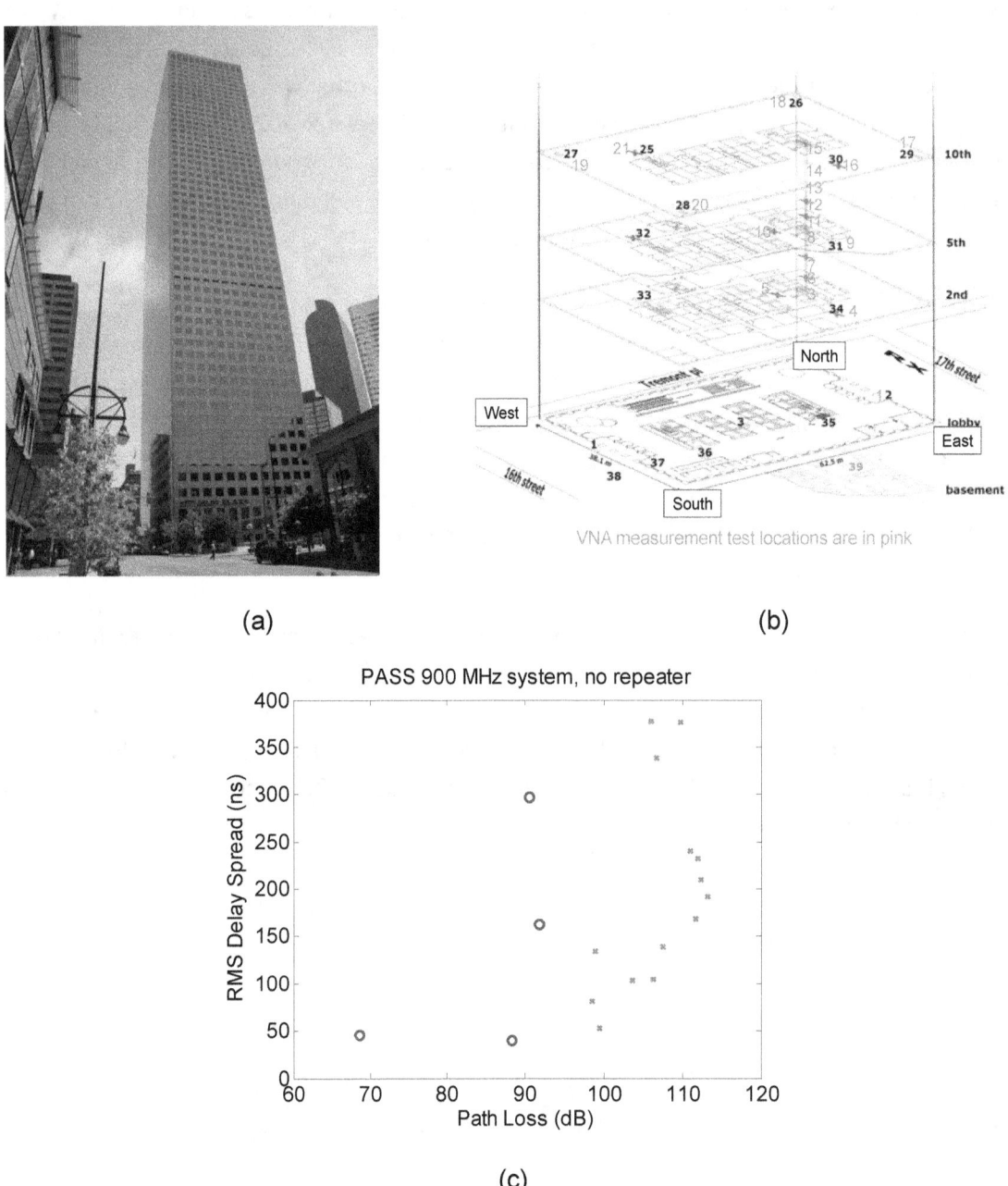

Figure 1: (a) 57-floor office building in downtown Denver, CO. (b) NIST researchers measured path loss and RMS delay spread at several locations on the lower ten floors of the building. (c) A plot of measured RMS delay spread vs. path loss, where the blue circles represent a successful RF-based PASS transmission and the red stars represent a failure [70].

In this and almost every case we studied, there was a direct correlation between an increasing path loss and failure of the RF-based PASS device to receive the base station's transmission. Conversely, failures can occur for any value of RMS delay spread. Thus, we conclude that it is necessary to rigorously test RF-based emergency equipment for performance under various attenuation conditions. However, most of the environments we studied had relatively short RMS delay spread values of 200 ns or less. It is possible that higher values of RMS delay spread may impede successful RF communications.

7.3.1 Measurement Results

To provide more detail on the types of radio-propagation environments that were tested in the NIST field measurements, we give a brief description of the type of environment, followed by any pertinent details on the measurements themselves and/or the resulting data.

The NIST-measured data are summarized in Figure 2. Figure 2(a) shows the average path loss, and the spread of values around the mean, while Figure 2(b) shows the RMS delay spread. In some cases, there was insufficient received signal level to calculate the RMS delay spread. The number of points used to generate each graph is not always identical. The standard uncertainty in the path loss measurements is approximately 5.5 dB, from [66]. See Ref. [70] for more information on these measurements.

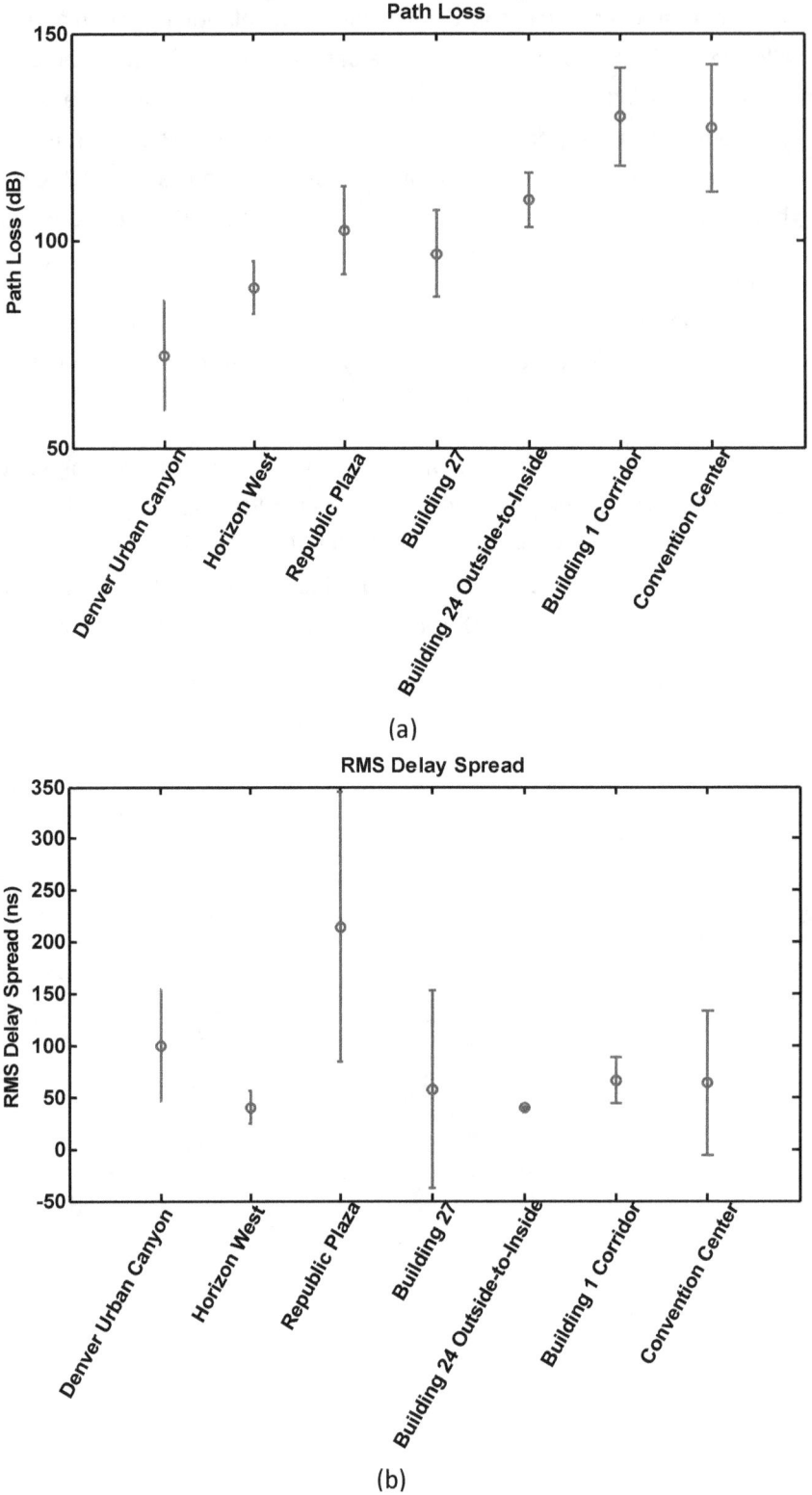

Figure 2: Radio-propagation channel characteristics measured several representative environments where emergency-responder wireless communications may be difficult. (a) Mean path loss. (b) Mean RMS delay spread. The bars represent the spread of measured data around the mean. The standard uncertainty in these measurements is approximately 5.5 dB.

Denver urban canyon – down street, around one corner: This was the only outdoor-to-outdoor environment studied. Measured path loss was between 45 dB and 90 dB, but the RMS delay spread was as high as 220 ns, one of the larger values measured. The RF-based PASS devices operated successfully in this environment except in two locations, both having moderate path loss and RMS delay spread.

Horizon West – 12-story apartment building, Floors 2 and 7: The path loss due to building penetration in this environment was not high, between 75 dB and 100 dB, and the RMS delay spread was less than 55 ns on Floor 2, and 80 ns on Floor 7. However, the RF-based PASS transmissions were generally not successfully received by the base station. A cell-phone base station located on the roof of the apartment building may have caused interference.

Republic Plaza – 57 story office building, Floors 1-10 including stairwell: The plots of RMS delay spread vs. path loss in Figure 1 clearly show that attenuation, not multipath, was the dominant channel impairment in this environment. When repeaters were used, reception was improved. The RMS delay spread values ranged from 50 ns to 400 ns, where 400 ns is the longest RMS delay spread in our set of measurements.

Note that the path loss values reported in Figure 1(c) (≈70 dB to 115 dB, whose values have been confirmed by comparison to single-frequency measurements made in [66]), are actually too small to cause the RF-based PASS to fail. In the NIST lab, the RF-based PASS devices failed to successfully transmit an alarm at an attenuation level between approximately 120 dB and 135 dB, depending on the device. For the results shown in Figure 1, we expect that the PASS base station antennas were not oriented properly to align with the remote units on the higher floors. Reorientation would be needed because the base station antennas have a null in the vertical direction. Consequently, the absolute values of path loss presented here do not correspond to the absolute path loss at which the PASS communication fails. However, the results do clearly show that, even for the relatively high multipath in this structure (up to 400 ns RMS delay spread), attenuation was the primary cause of failure, as opposed to multipath.

NIST Building 27, Boulder, Colorado – small main building connected by long subterranean tunnel to small back room: Attenuation values ranged from around 85 dB to 100 dB, and RMS delay spread was generally low, but in one case it jumped to 250 ns, probably because of multipath in the front building before the signal propagated down the hall to the receiver. PASS measurements were not made in this structure.

NIST Building 24, Boulder, Colorado – office/lab building with basement: Measured attenuation values ranged from 95 dB to 115 dB. The attenuation in the basement was higher than 115 dB, but we were unable to measure it due to the limited dynamic range of the VNA test set-up. The RMS delay spread values were low, only up to 50 ns for the locations we measured. From the limited data [70], we expect that attenuation was the dominant failure mechanism.

NIST Building 1, Boulder, Colorado – office/lab building with long, partly subterranean hallway: Measured attenuation values in this structure ranged from 100 dB to 140 dB and the RMS delay spread values were up to 100 ns. We were only able to acquire data at the locations nearest the transmitter because there was insufficient dynamic range to acquire path loss and RMS delay spread data. It is expected that the path loss is significantly higher deeper inside the building. Because of the short RMS delay spread and large value of attenuation, we expect that attenuation is the primary failure mechanism in this environment for the RF-based PASS devices.

Denver, Colorado Convention Center – main entry, one corridor, and downstairs: The measured attenuation in this large structure was between 80 dB and 150 dB, and the RMS delay spread up to 200 ns. However, our measurement instruments had insufficient dynamic range far into the building, so it is anticipated that the path loss could become much higher than this. RF-based PASS transmissions were received up to a path loss of approximately 130 dB.

7.3.2 RF Classification of Structures

Based on the measurements summarized above[††], we have developed suggested guidelines for classifying buildings in terms of RF attenuation and multipath. Table 13 summarizes the suggested RF classifications. Note that these are very broad classes. Other work [82] has attempted to provide finer levels of classification, but our goal here is to develop lab-based test methods that can be used for performance verification of emergency response wireless equipment such as LTS.

Most of the environments tested exhibited at least 50 dB of attenuation, created by the penetration of signals from outside to inside a structure (or vice versa) plus the free-space distance between transmitter and receiver. Only the outdoor urban canyon environment and the shallow Horizon West apartment building had maximum attenuation values less than 100 dB. It is expected that typical one- and two- family dwellings, small commercial buildings (such as small stores in strip malls and office buildings with exterior-facing offices) and small-to-moderate sized apartment buildings (in which all apartments have an exterior wall) would provide an environment where the total signal attenuation is less than 100 dB. We classify this type of structure as "low attenuation." With current (2010) technology, it appears that an individual RF-linked LTS (with no repeater) could operate successfully in these environments, unless external radio interference is experienced, as it was during the Horizon West apartment building measurements.

Most of the environments we studied had maximum attenuation values between 100 dB and 150 dB, which we classify as "medium attenuation." It is expected that moderate-sized structures such as small hospitals, moderate-sized and tall commercial buildings, office buildings, and apartment buildings would provide an environment with attenuation between 100 dB and 150 dB. With current RF technology, the use of a repeater can often overcome this level of attenuation.

Very large structures and those with subterranean floors can be expected to provide attenuation greater than 150 dB, which we classify as "high attenuation." NIST Buildings 24 and 1 and the Denver

---

[††] The measurements described above were made with an incident command station located directly outside the building, communicating to a firefighter inside the building, to represent a typical incident response scenario. This "point to point" communication scenario is different than using cell based or trunked based radio systems, which operate with base station towers, typically located a significant distance from the incident scene.

Convention Center had these high levels of attenuation. It is expected that multiple repeaters would need to be used in such environments, using current RF-linked LTS technology.

The largest value of RMS delay spread in our measurements occurred in the Republic Plaza building, around 400 ns. Based on our measurements of RF-based PASS, it is expected that most LTS can operate successfully in environments having this value of RMS delay spread. However, it is expected that in a large, highly reflective environment, multipath may become a more significant problem.

As noted above, the RMS delay spread in the environments we studied did not exceed 400 ns. As a consequence, our classification focuses on attenuation rather than multipath. The approach taken here is consistent with model building code and NFPA fire resistance building construction categorization [1][2], in that the building construction categorization is performed based on the combustibility and fire resistance of the building, while the RF classification is based on the attenuation and multipath characteristics of the building and its contents. It is important to note that the building construction classifications, however, do not directly correlate to the RF classifications. A summary of our proposed RF classification is provided in the table below.

Table 13. Classification of Structures in Terms of RF Characteristics Due to Building Signal Penetration.

Classification	Attenuation (dB)/ Multipath (RMS Delay Spread in ns)	Typical Structures	Current RF Technology
Low loss	Less than 100/ Less than 400	Houses, small buildings with exterior-facing rooms	Single unit
Medium loss	100 to 150/ Less than 400	Moderate-sized and tall structures with some interior rooms	With repeater
High loss	Over 150/ Less than 400	Very large structures and those with subterranean floors	Multiple repeaters
High loss, highly reflective	Over 150 dB/ ~1000	Very large structures with open floor plans and reflective exterior surfaces. Outdoor urban environments.	Multiple repeaters

7.4   Summary

NIST test results to date indicate that lab-based experiments providing methods for testing the RF communication link for emergency locator beacons in a controlled attenuation and multipath environment would predict device performance in many real-world firefighter environments in the absence of external RF interference (RF interference is covered in separate, but related testing). Tests utilizing various values of attenuation and multipath could be used to verify device performance in environments having the RF classifications listed in the table above.

Additional field tests and analysis should be conducted to determine the level of multipath in highly reflective environments such as factories, warehouses, and cluttered outdoor environments, and lab-based tests should be developed if it is found that these environments affect RF-linked LTS performance. In addition, it is critical that interference tests be developed, because of the potential for interference to interrupt the RF-linked LTS transmission, even when the size and composition of the environment should present no problem to successful reception.

Finally, the data collected on RF attenuation and multipath environments provides insight into key RF conditions that must be accounted for during assessment of the RF technology associated with LTS. The scenarios particular to RF communications reflect the knowledge gained from the measurement and analysis results described above. The inclusion of RF interference discussed in the next section must also be considered in the RF communication scenarios.

8   Radio-Frequency (RF) Interference Sources for Locator Tracking Devices

8.1   Introduction

In addition to channel impairments such as attenuation and multipath, the success or failure of an RF transmission often depends on whether external RF interference is present and whether the RF-based system is able to operate effectively in the presence of such interference. RF interference may arise from electromagnetic sources of energy in the RF spectrum. Figure 3 displays a number of applications that operate within the RF spectrum, including current LTS technologies. As suggested in this figure, the vast majority of RF interference sources for LTS will occur within the UHF and microwave frequency range, from around 400 MHz to 6 GHz. Such RF sources are ubiquitous in the environment and include intentional transmitters such as voice radios, wireless LANs, GPS signals, RFID systems, RF ranging systems, microwave ovens, medical and industrial equipment, and broadcast radio and television. Unintentional transmitters such as relays, switches, motors, and fluorescent lights can also interfere with RF transmissions.

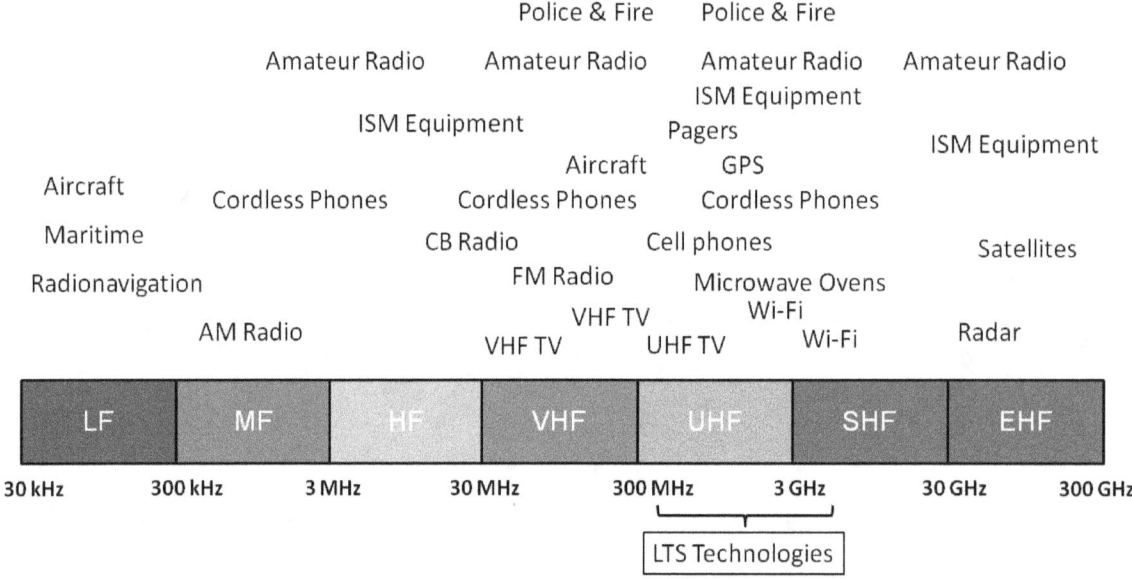

Figure 3. RF spectrum with selected applications. LF = Low Frequency, MF = Medium Frequency, HF = High Frequency, VHF = Very High Frequency, UHF = Ultra High Frequency, SHF = Super High Frequency, EHF = Extremely High Frequency.

Any component of an LTS that relies on RF signals is susceptible to RF interference. For example, a precision LTS based on RFID technology may be susceptible to RF interference corrupting the ability of the system to accurately interrogate the RF tags located either throughout the structure or on fire department personnel. While any RF-based system will fail in the presence of strong enough electromagnetic interference, it is critical that LTS technologies that rely on RF transmission for location and/or communication activities be capable of operating in the RF environment at the incident, or alert responders of their failure to reliably transmit. Therefore, testing scenarios should include representative sources of RF interference. An appropriate RF scenario will include potential RF sources already present within the building as well as sources brought in by emergency services.

8.2   Background

Technically, electromagnetic interference may be due to radiation, conduction, magnetic induction and electrostatic discharge. In any typical emergency response scenario, a large number of potential RF interference sources are likely to exist. As examples, the complexity of the RF environment is illustrated in an assessment of potential RF interference sources in a hospital or medical environment [51] and in a summary of devices that generate RF fields in general and workplace environments [85]. Note that while ref. [85] is focused on the health aspects of RF fields, the sources of interference are applicable to RF communications. Along with these potential sources of interference, the RF-based equipment used in an emergency response effort is likely to introduce additional RF interference. Voice radios (5 W) and repeaters (> 20 W) are examples of RF-based equipment with the potential to interfere with an LTS.

While any RF interference mechanism can potentially interfere with locator technology, interference from RF and wireless systems operating directly in or sufficiently near the bands used by the LTS are typically the most disruptive. Such systems include virtually any radio or wireless device such as a cell phone, hand-held radio, wireless access point, etc. RF interference effects between two wireless devices can take a variety of forms, such as transmitter non-harmonic broad-band noise, harmonic distortion, and desensitization, where a strong signal overpowers sensitive components in the receiver [86]. But by far the most common interference mechanism occurs when multiple signals that are transmitted in the same frequency band arrive at the receive antenna, are mixed together and downconverted, "confusing" the receiver when it tries to demodulate the signal.

In general, the disruptive impacts of RF interference depend on frequency, power, and signal modulation, including techniques that use time-sharing of transmissions to reduce interference. The physical proximity between the interference source and the LTS is also a key factor, and one way to minimize interference is to increase the separation distance between the LTS and the interfering source. RF power typically decreases at a rate between $1/R^{1.5}$ to $1/R^6$ in buildings, where a decrease proportional to $1/R^2$ is equivalent to a free-space far-field environment [87]. R is the separation distance between source and receive points or in this case, the distance between the interference source and the locator device. While FCC (commercial) and NTIA (government) regulations are designed to limit the interference between devices by limiting their output power, it is impossible to restrict the allowable radiated power to a level that eliminates any possible interference because practical physical separation distances may be closer than those specified in the FCC/NTIA regulations. For example, a firefighter with a locator device may be in need of assistance while positioned in near proximity to an RFID reader.

While the physical distance between RF devices is important, the separation of transmissions in frequency is also a useful way to mitigate interference. Frequency bands are assigned for use by the FCC for specific services such as public-safety, cellular telephony, and industrial, scientific and medical (ISM) applications. Basically, a transmitted signal must decrease below a certain level (for example, -100 dBm), outside of its assigned frequency band. The intent is to minimize RF interference to wireless devices operating in other frequency bands. Another key consideration is that, for efficient use of limited spectral resources, most frequency bands are shared, either by systems with a license to operate within that band, or, for unlicensed frequency bands, by systems that choose to operate with the prescribed FCC limits in those bands (ISM frequency bands are unlicensed). A comparison of radio power in the heavily used 2.4 GHz ISM band to that in the protected GPS L1 (Global Positioning System) and UNI-S (electronic news gathering and space communications) bands in the San Francisco Bay area in 2003 demonstrates the noisiness of the ISM band, even in rural sites [88]. Consequently, radio transmissions in the ISM bands use high-level modulation formats and transmission protocols to minimize interference.

Modulation is the imposed variation in waveform properties that allows the RF signal to transmit information. The type of modulation used directly affects how the transmitted RF energy occupies the frequency spectrum. Analog forms of modulation are most appropriate for low data rate applications. They are often used for licensed, narrowband public-safety applications, such as voice radios. Some LTS use analog modulation as well. However, to achieve a higher data rate while maintaining spectral efficiency, many wireless systems now use digital modulation. Digital systems are often less susceptible to RF interference because error-checking codes can be incorporated into the bit pattern itself, allowing the receiver to distinguish the desired signal from the interferer

From an RF interference standpoint, another important factor is the "channel access method," that is, the method whereby digital channel access methods enable a single transmitter to serve multiple users. Access schemes are key in minimizing interference by using frequency sharing (frequency division multiple access (FDMA)), time sharing (time division multiple access (TDMA)), or by reducing the energy emitted at a specific frequency by spreading the signal over a band of frequencies (code division multiple access (CDMA) and direct sequence spread spectrum (DSSS)). Newer access schemes use multiple sharing methods, both to maximize spectral efficiency and to minimize interference. These include frequency-hopping spread spectrum (FHSS) and orthogonal frequency division multiplexing (OFDM). To enable multiple transmitters located nearby to share frequencies within the same band, dynamic channel allocation (DCA) or dynamic frequency selection (DFS) may be used. These techniques rely on time sharing, combined with systems that can monitor interference and change their frequency of operation dynamically.

8.3 Approach to Interference Testing

Testing for all the different possible modulation techniques and channel access methods would represent an extensive undertaking. A more practical and feasible approach is to test for a combination of the expected RF technology in the building and the RF technology supporting the incident response, with the specific locator technology taken into consideration. For example, if the locator technology uses a 1 W, 900 MHz CDMA radio, disruptive RF interference from a 100 mW, 2.4 GHz FDMA system is not very likely. However, a 5 W, 905 MHz FDMA radio located within a couple of meters could potentially disrupt the LTS.

In order to provide a level of confidence in the interference tests, the first step is to develop a reasonably complete list of potential RF sources. Knowledge of the potential interference sources allows one to ascertain their key features from FCC or NTIA rules. Different rules apply to systems that are mobile versus fixed, operate indoors versus outdoors, operate in various frequency bands, with various transmitted power levels (or allowable field strengths), and with various modulation and channel access methods.

For this technical note, a list of potential RF interference sources in or near frequency bands that have been proposed for the LTS has been generated, based on FCC regulations. Some of the permitted emitters in each band used by current LTS technology are shown in Figure 4. More details, including limitations on power and field strength, are presented in the tables in the Appendix. Emphasis has been placed on sources that are most likely to share the frequency range of the LTS, although high-power sources outside each band are also included. These tables are not intended to be comprehensive, but they contain many sources that could be encountered at the scene of an emergency within a building. The tables should be used with caution. The geographical differences incorporated within the regulations are not included (e.g. urban/rural, east/west, near airports). Depending on the jurisdiction, the FCC may allow waivers to permit higher transmitter strength in cases where the limitations do not satisfy the operational needs, such as for public safety communications where attenuation of the signal must be overcome. Alternatively, the power levels of actual sources may be much lower than those allowed by the FCC or NTIA, especially for powerful transmitters such as broadcasting stations. In addition, the regulations themselves are in flux, changing in time to meet the evolving needs of the community.

From the master lists, the appropriate RF interference sources for scenarios to challenge an LTS may be selected based on a combination of the RF characteristics of the LTS and the building type (i.e., hospital versus suburban home). The selection of interfering sources should be prioritized based on both the anticipated power level and operational frequency band of the LTS. It is important to consider interfering sources not located in the actual operational band but close enough in both frequency and physical proximity to potentially cause disruption. The next several paragraphs point out some additional factors that should be considered in the selection and development of appropriate RF communication scenarios for LTS evaluation.

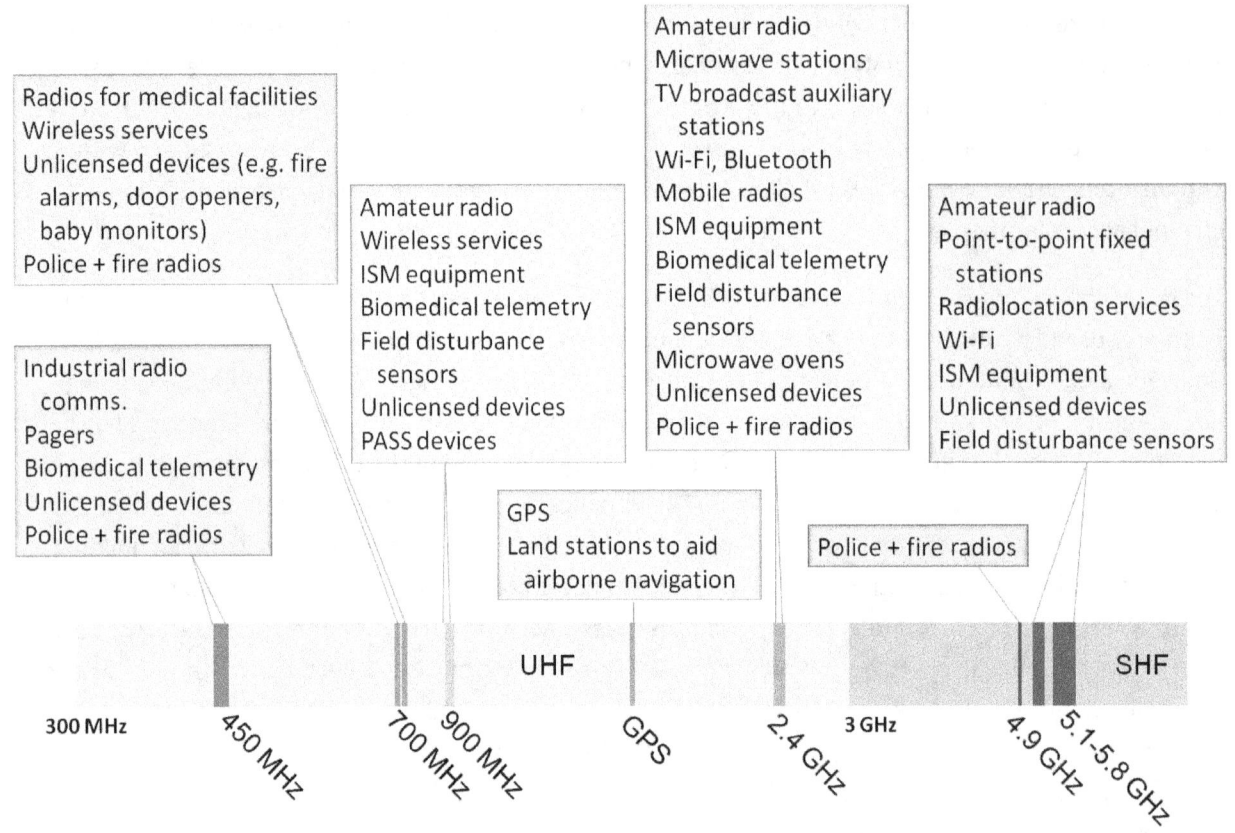

Figure 4. Potential RF interference sources within the frequency bands of current LTS technologies

Antenna gain patterns add to the complexity of the problem, since the pattern may create a significantly stronger interference signal at different, equally distant physical locations. For example, a narrow beam antenna can generate a 20 dB to 30 dB (100x to 1000x) difference in interference power level for a device located in the center of the main beam compared to a device located near a null point in the antenna pattern, even though the physical separation distance from the interferer is the same for both devices. In order to mitigate this type of problem, the FCC limits typically are specified in terms of an Effective Radiated Power (ERP) or Effective Isotropic Radiated Power (EIRP). EIRP includes the gain of the antenna along with the maximum permitted transmitted power. Finally, when the interference source and the LTS reside in each other's "near field," the electromagnetic interactions, based, for example, on mutual coupling between antennas, become exceedingly complex. This makes prediction of RF interference behavior extremely challenging.

In the FCC and NTIA rules, as well as in technical standards, there are two common ways of representing RF energy: as transmitted power (typically a maximum or minimum), or as the field strength at a given distance (e.g., millivolts/meter (mV/m) at a 3 meter distance). The choice of representation depends mainly on the type of analysis, where transmitted power is generally used when discussing communication and radar systems and field strength is used in electromagnetic compatibility (EMC) and electromagnetic interference (EMI) problems. An interesting discussion of the potential for interference based on devices using the two different specifications can be found in [89].

Often power limits are set by human exposure limits rather than the FCC allowable maximum. This is especially true for high-power communications towers located near population, for example, on an apartment roof top. Another important consideration is that the field strength from a tower mounted on a neighboring building may be higher than the field strength from one mounted on the roof, due to the directionality of the antenna. For a good introduction relevant to EMC/EMI problems involving antennas, conversion between power and field strength, as well as other electromagnetic propagation mechanisms such as reflections off surfaces and multipath contributions, see ref. [90], Chapter 5. However, as pointed out in [91], the notion of wireless coexistence is also well-aligned with the challenges covering the possible RF interference to LTS devices.

GPS-based systems differ from other LTS designs in that satellite signals are used for determining location. The GPS signal is very weak after reaching the Earth from a 50 W source on satellites orbiting at 32 000 km (20 000 miles) [88]. GPS receivers are thus very sensitive and may be susceptible to other RF activity at the emergency scene, even from frequencies well outside of the GPS band. In addition, the need for the receiver to lock to signals from at least three satellites for a 2D position and four satellites to add the vertical dimension, as well as errors due to reflections (multipath) from tall buildings, make the use of GPS problematic in urban environments.

9   Challenging Scenarios for LTS

In Section 5, LTS technologies were described in general, and in terms of their strengths and weaknesses. In this section, scenarios will be developed that challenge the weaknesses of the LTS technologies. The development of the scenarios in this section is primarily based on the weaknesses of the LTS technologies, and also considers the strategies or tactics employed by the fire service, but does not consider conditions that may affect communication with the IC post, which is discussed separately. A basic assumption for all the scenarios is that visibility within the structures is zero.

Basic dead reckoning is not included in this discussion because it is not used by the fire service per se, however, inertial technology is included. These scenarios depict challenges to each LTS technology individually but if strong similarities are found within the individual scenarios, they can be combined and applied to multiple technologies. Recall that it is assumed that the size, weight, ruggedness, cost, and user interface of each LTS technology is acceptable to the emergency response community.

9.1   Audible Alarms

The weaknesses associated with audible alarms used for LTS purposes are centered around the ability of a rescue team to 1) hear the alarm sound above the ambient noise at the emergency event, 2) recognize the alarm sound as a trigger for urgent action, and 3) use the alarm sound to locate the firefighter in distress. A scenario that challenges these three potential weaknesses would therefore include a high level of ambient noise, the presence of similar sounds, and one or more conditions that confound the homing properties of the alarm sound.

To expand this idea, consider a fire in a building that has many narrow passages, *e.g.*, an apartment building or a storefront that contains several retail businesses having some common facilities. The building is located in a high density area and is surrounded by tall buildings that reflect street noise. In this scenario, a fire occurs and responders from several different organizations arrive. The audible alarms used by each organization emit a different sound. The ambient noise level within the burning structure is very high due to firefighting operations, equipment, and noise coming from outside the structure that has been reflected from the surrounding structures. One of the firefighters goes missing at some point during the fire attack but before the fire has been contained. This firefighter mistakenly travels deeper into the building while trying to find the way out. Eventually the firefighter stops moving and the audible alarm activates, but not until the firefighter is separated from the others.

For a rescue operation to be successful in this scenario, the audible alarm sound must be loud enough to be heard over the noisy ambient sound environment, it must also be recognized by the rescuers as an audible alarm and not be confused with any other sound, and it must guide rescuers through passages that may have changed during the course of the fire.

9.2   Inertial

As described in the previous section, inertial technology uses accelerometers and gyroscopes, coupled with error correction algorithms and knowledge of human motion, to establish the location of a firefighter. The major weaknesses of inertial technology relate to the ability of the error correction algorithms to compensate for natural and manmade environmental interference, the presence or absence of known points with which to reset the error correction calculations, and the fidelity of the method of integrating firefighter movements.

In this scenario, a fire has started in a wood frame single family home having two full floors and a loft or mezzanine above ground. It is uncertain whether the home is inhabited at the time so search teams are deployed in the upper floors while the fire is being attacked below. The stairway is roughly in the center of the house near the kitchen. One search team goes up to the second full floor while the other team searches the ground floor and then moves on to the mezzanine area, where they are met by the team coming down from the second floor. With the searches complete, both teams return to the ground floor and exit the building.

The fire has extended into the space under the stairwell while the searches were underway and has weakened the stairs. As the two teams were retreating the two last firefighters in the loft / mezzanine area encounter trouble. One of them, a firefighter that had been searching the ground floor, plunges through the weakened stairs and down into the basement. The remaining firefighter feels the floor of the mezzanine starting to buckle, tries the stairway up to the second floor, finds that it is no longer passable, and then moves to a more secure section of the mezzanine to wait for rescue. The firefighter that fell through the stairs to the basement is now in the fire room with the fire blocking escape. The RIC must cut through two walls to rescue the firefighter.

The challenges for the LTS in this scenario include determining the proper floor of the firefighter in the mezzanine, which is not at a full floor's height above the ground floor, and locating the downed firefighter in the basement. This firefighter had been searching in the vicinity of the kitchen before going upstairs. The RIC is using an axe and chain saw to chop through walls to rescue the firefighter. Potentially, the location of both the firefighter and the RIC could be significantly uncertain.

9.3 Ultrasonic Waves

Ultrasonic waves require a clear path within which to propagate a signal. Ordinarily, there is a hose line nearby that, in addition to supplying water, tends to keep doors open and thus ensures that an ultrasonic signal can make its way out of the structure. Gases expand as they heat up, building pressure if they are enclosed, or flowing toward lower pressure regions if they are not enclosed. In either case, hot gases can shatter windows and slam doors shut. Building materials also change during the course of a fire, sometimes opening up new passageways and sometimes collapsing existing spaces.

In this scenario, a fire has started in a single family home with a basement. The home has one floor above the basement, but it has undergone numerous additions since its original construction and now has several bedrooms joined by a hallway on one side of the house and several other bedrooms joined by another hallway on the other side of the house, with a den, living room, kitchen, and entry way between. The house is compartmentalized, with doors separating both blocks of bedrooms and their respective hallways from the central living area.

The fire started in the kitchen, has extended into the attic/ceiling spaces, and is spreading into other rooms from above. During the search for inhabitants, one of the firefighters is struck by a roof rafter and is trapped under it in a bedroom. There is an air vent for the central heating system near the firefighter that leads to the air handling unit in the basement. The door to the bedroom and the hallway are both open but there is no hose line present to hold them open. As ventilation operations are conducted and the fire moves into and out of various regions of the house, access to the bedrooms changes as well.

The LTS must be capable of reliably locating the downed firefighter in spite of changing conditions. If one or more doors slam shut, other openings may appear and may mislead the RIC or lead them to an air vent or the air handling unit in the basement instead of directing them to the firefighter.

9.4 GPS

Since GPS requires communication with several satellites, it has very limited use inside structures unless communication can be established.

If GPS is used to establish one or more known points at the initialization of an LTS, or for periodic position updates, then a scenario that stresses the system would be the operation of the unit in a dense urban setting (*i.e.*, tall buildings), with multiple public-safety entities using radios at the scene. The ability to initially calibrate an LTS with a GPS signal may be impaired if too much RF noise is present in the vicinity of the IC post.

9.5 RFID

As mentioned previously, the ideal scenario for RFID technology includes a building that has RFID tags or readers integrated into the design and the local fire service has complementary readers or tags with which to interact with the building's system. Suppose a fire starts in such a building. This is a single story commercial building in which electronic items are manufactured. Several hundred people are typically present during working hours. In addition to the firefighter RFID system, there are other wireless systems that provide internet access, radio and cell phone signals, and environmental controls

The fire starts in a coffee break room and spreads into a copier room, eventually setting off smoke detectors and a water mist suppression system, which slows the progress of the fire but does not extinguish it. There is some confusion during the evacuation process and a few people are unaccounted for when the fire department arrives. Two search teams are dispatched into the building through separate entrances. One team searches the administrative part of the building which is comprised of offices, cubicles and numerous RF sources. The other team searches the shipping/receiving and storage areas near the break room and adjacent to the administration area. This is the direction the fire has taken. Incoming and outgoing packages are stored in shelving that inhibits the effectiveness of the water mist, so gas temperatures in parts of this area are approaching 200 °C. When the team's searches are complete and they have checked in at the IC post, one firefighter from each team is missing.

This scenario has two challenges: first, the administrative area has an RF environment that can potentially interfere with the RFID system; second, the harsh conditions in parts of the shipping/receiving/storage area may cause the tags and readers to malfunction. If RFID tags and/or readers are used as part of a hybrid LTS that is not integrated into a building, the same two challenges would still apply.

9.6   Ranging Radios, Ad Hoc Mesh Networks

Since ranging radios and ad hoc mesh networks operate in a similar manner and can be configured using virtually the same variety of RF signal characteristics, these two technologies will be combined in the following scenario.

A fire occurs in a large, multi-story single family residence situated at the top of a ridge. The building is long and narrow with its long dimension parallel to the ridge contours. Access to the building is limited to the side facing the street due to the steep hillside and the presence of terraces and retaining walls around the other sides of the building. It is not possible to position the antennas for the ranging radios anywhere other than in the street in front of the building. The windows facing away from the street, toward the valley below, have been coated with a reflective material, and the exterior wall facing the street is brick. A cell tower and other broadcast antennas are located a bit further along the ridge.

Upon arrival at the scene, one search team looks for the source of the fire and another team looks for victims. The teams are trained to deposit breadcrumbs whenever their LTS tells them they are about to venture outside the range of the system. The team looking for the fire becomes engaged in fighting it and ignores the command to deploy breadcrumbs. The team looking for victims moves downward through several stories as they conduct their search. When they find a number of victims, they start looking for the quickest way to get them out of the building. When the team with the victims opens a double door to the outside on the bottom floor of the house, fresh air rushes in and intensifies the fire on the floors above. At this point the other firefighters in the building make a hasty retreat. All except one firefighter, from the fire attack team, are accounted for when they check in with the IC. The stairs become inaccessible due to the fire following the fresh air blowing in from below, which destroys any breadcrumbs that may have been placed in the vicinity of the stairs.

Most of the conditions in this scenario are challenging to RF systems, whether they are RFID, RF ranging radios, or RF-based ad hoc mesh networks. The most basic challenges are multipath, signal attenuation, and interference from other electromagnetic sources. This is also true for RF communication from inside the structure to the IC post.

9.7   Scenario Summary

The scenarios presented in this section are imaginary; they were created to illustrate the types of environments and situations that challenge LTS technologies and are not the only examples that could be used for this purpose. Given the scenarios presented here as a starting point, a fire department that is contemplating the purchase of an LTS may consider applying this thought process to the types of structures and the situations that have led to the need for rescue of lost, caught, or trapped firefighters within their protection district or similar jurisdictions. Knowledge of the weaknesses of each type of LTS technology and familiarity with the specific needs of the fire department can help firefighters make an informed assessment of the LTS most likely to meet their requirements.

10   Radio-Frequency Scenarios

The first set of scenarios focused on the performance of the LTS technologies without specific concern for the communication between the LTS device inside a structure and the IC post outside the structure. In the scenarios below, consideration is given to some potential communication challenges that arise when using such LTS technologies. The communication aspects of LTS are critical to the overall success since the location and tracking information must be available at the IC post in order for the firefighters to take appropriate action.

The communication systems are generally expected to be standard RF radios in licensed and/or unlicensed frequency bands. Thus, the scenarios described here focus on the RF signal challenges such as propagation path loss, multipath, and interference. Earlier sections described key features in classifying buildings based on their RF properties of path loss, multipath, and interference. The discussion and tables of possible RF interference sources illustrated that the LTS system will likely need to operate and communicate in a "shared" RF environment. Below, the intent is to describe realistic scenarios that focus on key RF challenges to the LTS communication system.

10.1   Path Loss and Multipath

The first four scenarios focus on the attenuation and multipath the RF signal experiences while propagating from the firefighter back the base station located at the IC . (The base station is the device that communicates from IC to the firefighters, and displays the location/tracking information.) Multipath effects due to the composition of the structure where the firefighter is located are also included.

10.1.1 Low attenuation, low multipath

The least challenging condition for RF communication is when the firefighter is located in a typical, two-story, wood frame residential dwelling that includes a basement. This is also the most common environment in which firefighters run into trouble, as discussed in Section 4. In this case, path loss is typically less than 100 dB and a minimal amount of multipath is expected. Other structures that fall in the low attenuation category include strip malls, small apartment buildings, and small office buildings. A basement will generally cause the most signal attenuation, and the greatest source of multipath will likely be the either the kitchen or basement areas that contain multiple metal objects such as refrigerators, stoves, furnaces, or metal ducting. Adequate communication coverage must include the basement, in particular the effects of any metal boiler or furnace type structures, which could introduce multipath.

10.1.2 Medium attenuation, low multipath

Larger apartment and office complexes typically fall into the medium attenuation (100 dB to 150 dB) classification, mainly due to the size of the structure. A 20-story office building, without coated windows, and a single level underground parking garage represents a medium attenuation, low multipath environment.

A scenario to test the communication system has the firefighter walk though the various floors, transitioning between the floors via the stairwell. An interior stairwell can cause attenuation and multipath due to the structural steel, metal doors, and thick concrete used in typical construction. The attenuation is generally not severe at the lower floors, but at higher floors the combination of distance from the IC post and loss due to building materials can lead to path loss values of between 100 dB to 150 dB. The path covered by the firefighter should extend to the furthest point from the IC post. Another key place to check for coverage is throughout the stairwell.

10.1.3 High attenuation, low multipath

A large convention center, with floors below the street level, represents a high attenuation, low multipath environment. Depending on the frequency, the lower floors will experience attenuation of more than 150 dB. Similarly, a below ground parking garage can expect path losses of greater than 150 dB if the IC post is good distance from the firefighter.

In this scenario, the firefighter will again walk through the various floors, transitioning via the stairwell. The path must cover the high-attenuation areas such as the middle portion of the convention center and the below ground levels. Multipath effects will generally be most noticeable in the stairwell and hallways, particularly if the hallways are with lined finish panels made of reflective material. In this scenario, the firefighter does not walk deep into the mechanical rooms, as that represents a high multipath scenario covered below.

10.1.4 High attenuation, high multipath

A large form-factor building with a great deal of manufacturing equipment, such as an automobile plant, represents a high attenuation, high multipath RF environment. Note that a large hospital is likely to experience high attenuation, but a less complex multipath contribution. In both cases, the potential for in-building RF interference is quite high due to equipment, either manufacturing or medical. RF interference scenarios are discussed below.

In this scenario, the firefighter would move throughout the building in a path typically followed during an incident response. The path should include the factory floor so as to test the communication system in a complex multipath environment. If the building has metal walls and is a large structure, the path loss between the firefighter and the IC post located outside the structure will be high, likely greater than 150 dB. The metal walls will add to the complexity of the multipath environment as well.

10.2   RF Interference

The next four scenarios focus on potential RF interference problems. RF interference issues are broken out separately from attenuation and multipath because the different interference scenarios may apply to some or all of the previous scenarios. Figure 5 depicts several possible interference sources to LTS communications. The interference scenarios are broken down into those likely to be encountered by the firefighter in the building and those experienced at the base station located at the IC post. It is neither practical nor possible to test for all potential source of RF interference, but these scenarios are intended to test for the conditions that are most likely to disrupt LTS communication.

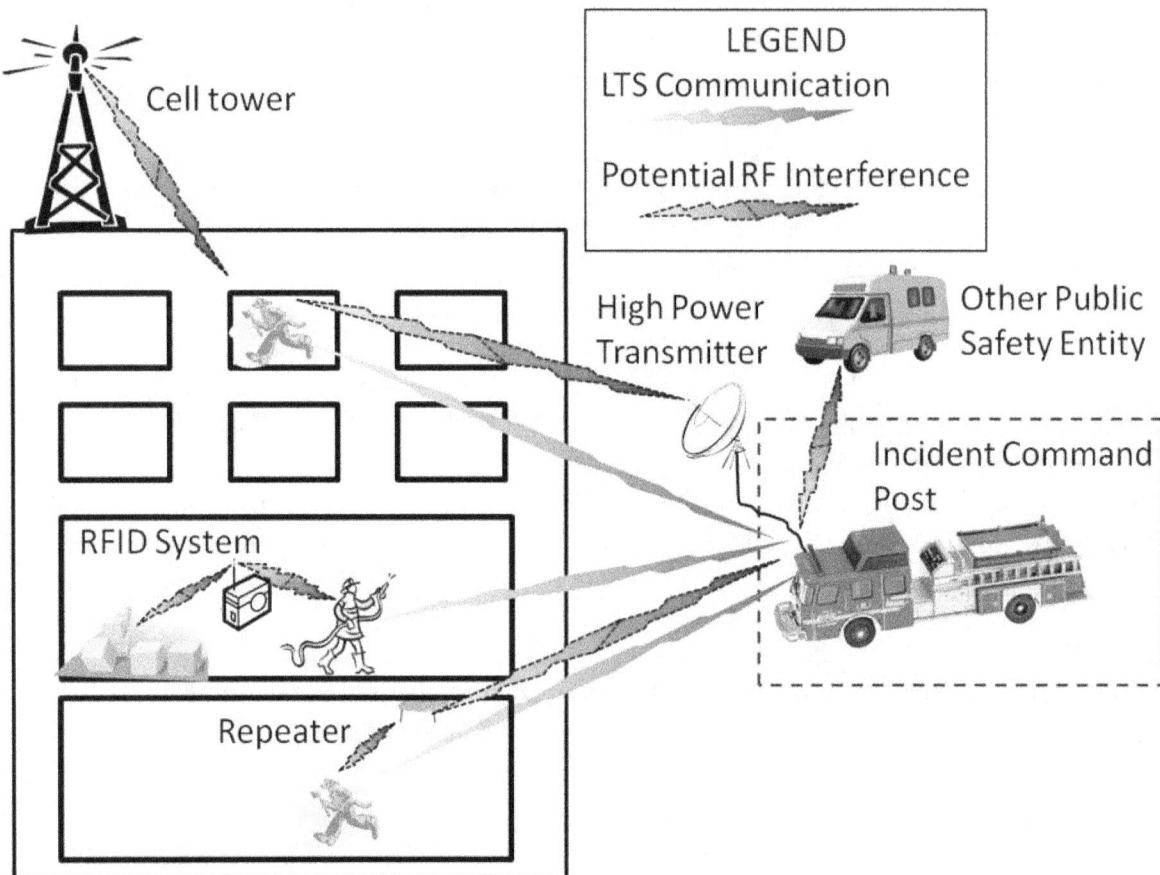

Figure 5. RF interference sources that can potentially disrupt or block LTS RF communications. The interference scenarios discussed below are designed to test the susceptibility of LTS communication system to these types of RF interference.

Tables A2 to A8 in the Appendix list potential in-band and out-of-band interference sources for the various communication bands likely to be used in LTS technology. The interference scenario that is chosen to test a specific LTS should use frequencies appropriate for that system's particular RF communication technology. In other words, a 900 MHz system should be tested for RF sources most likely to interfere with a 900 MHz system, and so forth. The information in Tables A2 to A8 help users focus on those sources that warrant the highest priority of consideration.

10.2.1 In-band interference at in-building nodes (firefighter or relay node)

The first type of RF interference test scenario consists of an intentional RF source operating within the same frequency band as the communication system of the LTS system. This is particularly important for LTS devices operating in unlicensed bands, such as the 900 MHz and 2.4 GHz ISM bands. Appendix Table A1 lists several potential frequency bands for the LTS communication technology, along with other possible in-band systems that may be operating within a building. The table clearly shows that both licensed and unlicensed bands may have other radio systems operating within that same band.

Although the licensed bands are coordinated to eliminate the use of a given frequency of operation (channel) by more than one user, interference is still possible if different jurisdictions participate in an incident response. In addition, if a particular jurisdiction used the same frequency band for both their voice radios and the communication system for the LTS technology, in-band interference is possible. Note that, in licensed bands, because each frequency band has multiple channels, the voice radio and LTS can operate in the same frequency band, but on different channels. While the channels may not directly interfere, the channels may be close enough in frequency that if the two systems get physically too close, interference can occur.

In this scenario, the firefighter would move in close proximity to an RF device radiating in the same band, but as part of a different communication system. For devices in the ISM bands, this means the firefighter moves near a device within that same ISM band, such as a wireless LAN access point or an RFID reader. Close proximity in this scenario is between 1 m and 2 m, where the intent is to replicate a firefighter walking by a device such as an RFID reader. The transmit power of the interference source should adhere to the regulations listed in Appendix Tables A2 to A8.

As appropriate for the intended deployment jurisdiction, the LTS communication system should be tested for immunity to a high-power in-band relay transmitter (otherwise known as a repeater). This scenario would then include the firefighter walking past the high-power, in-band node. Such repeaters are often installed in large buildings to assist in RF coverage. The minimum physical separation distance should be approximately 2 m. The transmit power of the repeater should be the highest power used within that band. For the purposes of this test, any relay device above 10 W would be considered high power.

10.2.2 Out-of-band interference at in-building node (firefighter or relay node)
In this scenario, the firefighter would move in close proximity to a device radiating in a different but nearby frequency band. This scenario is intended to capture the effects of a firefighter moving in the vicinity of a high-power radiator such as a cell-tower on the roof of the building or an adjacent building. A high-power out-of-band repeater can create a similar interference problem; however, the interference does not generally cover as large of a physical area within the building because repeaters usually operate at lower powers than do cell phone base stations, for example.

The specific frequency of operation for the RF interference source should be chosen to represent an out-of-band high-power transmitter close to the LTS communication band. For example, Appendix Table A4 lists the 800 MHz cellular telephone service as a near-band radiator for LTS communication systems in the 900 MHz band. To test for potential interference, the firefighter would walk to the top floor of the building containing the cellular base station antenna. Oftentimes the power transmitted directly below a cellular base station antenna is stronger at a horizontal distance from the antenna than directly below it, due to the antenna's radiation pattern. If the building containing the antenna is located near other buildings of comparable height, the firefighter should also walk through the top couple of floors of the adjacent building.

10.2.3 In-band interference at base station (located at IC post)

RF interference at the IC post can be particularly problematic due to the increasing use of wireless technologies. For example, in Appendix Table A2 there are public-safety radio systems that may be present at the IC post having EIRP values of 20 W or more. For this scenario, the antennas for the LTS base station and another in-band, high-power public-safety radio system are located as they would be at an emergency response scene. The physical separation between the two antennas should represent the minimum separation expected at an incident and the RF interference source should operate at its maximum expected power. It is anticipated that the locations of these antennas are fixed during the emergency response event. The operation of the LTS in the presence of the in-band, high-power radio system is tested.

A variant of this scenario tests for the potential interference from a lower power communication system located at the IC post that uses the same wireless technology as the LTS. In this scenario, both systems use the same communication technology, such as IEEE 802.11b or g in the 2.4 GHz ISM band. If the IC post is using a wireless LAN in the same band as the LTS technology, the wireless LAN can potentially interfere with the LTS base station. Appendix Table A6 lists a number of potential in-band RF interference sources in the very popular 2.4 GHz ISM band, both mobile and fixed. The mobile devices are problematic because they can inadvertently be used in transmit mode near the LTS base station. The scenario should test LTS base station immunity from mobile public safety wireless devices operating within 2 m to 3 m of the base station.

10.2.4 Out-of-band interference at base station (located at IC post)

This scenario focuses on potential out-of-band interference problems at the IC post due to a high-power transmitter or repeater brought to the incident to support RF voice communication to firefighters or other public-safety entities. This high-power transmitter is used to overcome building attenuation, as discussed in the attenuation and multipath scenarios. Unfortunately, such high-power transmitters can significantly interfere with the LTS base station. For example, Appendix Table A2 lists a possible near-band interference source from a base station operating in the 470 MHz band that can potentially interfere with a 465 MHz LTS communication system. This means that if a high-power source, for example a 35 W base-station transmitter, is used to improve public-safety voice radio communication, the sensitive receive electronics in the LTS base station may become overloaded, and unable to receive the relatively weak signal emanating from an LTS device on the firefighter within the building.

This particular scenario points out a key difference between two solutions for solving inadequate RF coverage within a building. In the case of the 35 W base station, one solution is to increase the transmitted power level (say for the voice radio system) to overcome attenuation. Note that the radio channel is not necessarily reciprocal in this case since the firefighter's unit cannot also transmit at 35 W. Because of the potential for the 35 W transmitter to overload the LTS base station's receiver, a proportionally less-sensitive receiver system would need to be used at the LTS base station, reducing the effectiveness of the LTS. A second approach is to use a multi-hop system, where the firefighter is connected through a series of wireless devices, which may include other radios of other firefighters. Each of these devices transmits comparable power levels to create a network of wireless devices. Mesh networks utilize this principle for RF connectivity in large areas. The difficulty with this approach is that a relatively low-power mesh network can experience significant RF interference if it is operating on a frequency near a high-power transmitter. Thus, any LTS solution based on a mesh network should be tested for its reliability in the presence of high-power transmitters and repeaters.

11   Conclusions

In order to provide confidence to the users and purchasers of LTS, performance metrics and standards must be developed and published. As part of this effort, firefighter structural scenarios and RF scenarios (propagation and interference) representing the environments in which firefighters operate were developed.

To assist in determining the scenarios where firefighter LTS technologies may be of the most benefit, a brief history of representative fire incidents in which firefighters were fatally injured, and whom a firefighter LTS may have made a difference, were presented along with firefighter injury and fatality statistics. The statistics indicate that firefighters fatally injured while lost in structures occurred most often in residential one and two family homes and in buildings of unprotected wood frame construction. In addition, it was found that the average yearly number of fireground injuries is greatest in one and two family dwellings, while the average yearly injury rate per 100 fires is greatest in industrial, utility, and manufacturing occupancies. The statistics also provide a first-order estimate of the numbers of firefighter injuries and deaths that may be prevented if effective firefighter LTS were available and used.

As discussed in this technical note, a single (currently available) LTS technology is unlikely to perform reliably within the wide range of environments in which the fire service routinely operates. Each LTS technology has its particular set of performance weaknesses. Audible alarms, for example, must be heard, recognized, and responded to, and then must direct a rescue team to the sound source in a noisy environment, with the LTS performance dependent on the quality of the alarm sound. The performance of inertial technologies depends on the strength of the error correction algorithms, resistance to magnetic interference, and the degree to which human movement is characterized. Ultrasonic technologies suffer from signal attenuation due to lack of open space through which the signal can propagate, and from multipath effects. Depending on the system configuration, RF-based LTS technologies are potentially vulnerable to performance degradation due to lack of line of sight communication among system components, signal attenuation due to building construction materials, multipath effects, and interference from the electromagnetic environment. Other, secondary situations may exist that also contribute to LTS reliability or signal quality degradation, such as changing fire conditions or firefighters that don't deploy breadcrumbs properly. Given the challenges posed by both structural and RF scenarios, it is likely that an effective locator tracking system will incorporate two or more technologies.

The RF scenarios provided in this technical note point out key features required in realistic tests of RF propagation and interference conditions for LTS communications in the emergency response environment. From the RF signal perspective, basements and very large buildings often cause significant signal attenuation, while mechanical rooms and factory floors can create complex multipath reflections. These conditions may disrupt the RF communication channel, ultimately leading to a failure in LTS communication capabilities. In addition, even though RF wireless communication devices may adhere to government regulations and/or industry standards, the potential for RF interference still exists. Interference from other RF sources, such as RF relays and high-power transmitters, have a high potential to disrupt the communications of an LTS. The increasing use of wireless devices and technology for a variety of applications both within buildings and by public-safety entities raises the likelihood of RF interference impacts on LTS technology. The overall operation of LTS technology requires communication with the IC post and/or rapid intervention crew (RIC), which is typically performed with wireless communication technology. Thus, appropriate RF communications testing is needed to instill confidence in the dependability of LTS technology to support firefighters.

Finally, the structural and radio frequency scenarios were developed in order to provide firefighters and other stakeholders guidance in selecting and developing firefighter locator tracking systems. While LTS technologies may not be effective in all types of scenarios, fire departments and local jurisdictions may find that there are technologies that work well in their community. Some technologies, for example, may operate satisfactorily in buildings and occupancies that do not pose a serious RF challenge, such as unprotected wood framed single family homes. Fire departments that are contemplating the purchase of an LTS must consider the types of structures, occupancies, and situations that have led to the need for rescue of lost, caught, or trapped firefighters and develop scenarios that apply to the protected structures in their communities. Knowledge of the weaknesses of each type of LTS technology and familiarity with the specific needs of the fire department can help firefighters make informed decisions about the LTS most likely to meet their requirements.

12   Future Work

There is great interest in LTS for firefighters and other types of emergency responders. Due to this interest, research, development, and standards activities are active areas of work. During the development of this technical note, several areas of potential future work have been identified.

In the future, it would be valuable to have more data to determine the number of fire incidents where firefighters became disoriented or lost in a structure. One potential method would be to examine all of the historical firefighter fatality narratives in the NFPA FIDO system. These narratives would include incidents where disorientation may not have been classified as the first in the chain of events leading to the fatal injuries. Additional information on firefighter injuries related to becoming disoriented or lost in structures would also be valuable for determining where, and under what circumstances, firefighter locators could have the most impact.

The potential for RF interference is increasing as usage of wireless communication technology increases in public safety applications, as well as in the general public. The confidence in RF-based solutions to support the firefighter must include a thorough understanding of the potential sources of RF interference that a firefighter will encounter at an incident. Future work should measure and attempt to rigorously quantify the potential RF interference sources in frequency bands of particular interest to firefighters and other public-safety entities across a range of building types and emergency response events. In addition, as mesh networks are used to overcome the high attenuation problems encountered in large or hardened building as well as the below ground portion of structures, the mesh network's immunity to RF interference must be tested as a complete communication system rather than a single-node (point-to-point) interference problem.

One of the assumptions in this technical note is that the LTS equipment maintains the ability to operate while exposed to the conditions present in rough-duty firefighting environments. Future study is needed, however, to determine the ability of RF systems to communicate through firefighting environments that include fire, smoke, hot gases, dusts, aerosols, and water sprays.

Finally, the complexity of LTS technology coupled with the wide-range of building structures make evaluating LTS solutions a daunting task. Development of laboratory –type testing for both the locator/tracking and communication components of LTS would support test repeatability. The laboratory test would provide rigor and repeatability, and would help ensure confidence in LTS solutions.

Acknowledgements

The Science and Technology Directorate of the U.S. Department of Homeland Security sponsored in part the production of this material under Interagency Agreement IAA # HSHQDC-10-X-00408 with the National Institute of Standards and Technology (NIST). The National Institute of Standards and Technology funded this work as part of the Advanced Fire Service Technologies Program of the Engineering Laboratory. The authors would like to thank Jacob Healy of the NIST Electromagnetics Division for his contributions to the RF building classification section of this report, and Dr. Rita Fahy of the National Fire Protection Association, Fire Analysis and Research Division, for providing fatality data for firefighters lost in structures.

Appendix - RF Interference Sources in Specific Frequency Bands

The set of frequency bands that have been proposed for LTS use is listed in Table A1. In each of these bands, the FCC and NTIA permit the operation of certain RF transmitters, according to the Table of Frequency Allocations [33]. Tables A2 to A8 in this Appendix present radiative RF sources for each band that may be present at the site of the emergency and that transmit within the band or at nearby frequencies as determined from a review of FCC and NTIA regulations [34][37]. The sources are organized by placement inside or outside the building or as part of the emergency response, and by degree of mobility (mobile/portable devices versus fixed stations). On-site fixed sources include the possibility of antennas mounted on the roof.

Various measures are used to describe source strength limitations in the FCC and NTIA regulations. The transmitter output power (TOP) is the measurement of the power output of the transmitter itself without consideration of the antenna gain. The peak envelope power (PEP) is the average power supplied to the antenna transmission line by the transmitter during one RF cycle at the crest of the modulation envelope. The Effective Isotropic Radiated Power (EIRP) includes the gain of the antenna, and describes the power that the system would theoretically emit if it radiated equally in all directions. The Effective Radiated Power (ERP) is the transmitted power radiated in a specified direction and includes the effects of antenna gain. When the direction is not specified, the direction is assumed to be that of the maximum antenna gain. RF emissions may also be limited in terms of the maximum allowable field strength at a given distance. The tables in this appendix show the limitations as they are expressed in the regulations, although some values given in decibel measures such as dBm and dBW have been converted to watts.

The definition of transmitter types varies somewhat within the FCC regulations. Fixed stations are permanently set in a location, such as broadcasting stations and cellular towers. Mobile stations are defined by the FCC as stations that are intended to be used while in motion or during halts at unspecified points. Land stations are not intended to be used while in motion and include both fixed stations and base stations that are transported to the base of operations. Hand carried transmitters may be included in the definition of mobile devices. Portable devices are generally defined as hand-held units whose transmitters are intended to be held within 20 cm of the body of the user.

A large number of RF sources in today's environment are unlicensed. The FCC and NTIA do not require licensing for low power devices operating in certain frequency bands, including some of the LTS bands, as long as they meet certain requirements. This non-licensed equipment includes many devices that have become part of everyday life, including cell phones, cordless phones, Wi-Fi, infant monitors, electric door openers, biomedical telemetry, RFID, ISM (industrial, scientific, and medical) equipment such as microwave ovens, and many others. The limitations on intentional and unintentional RF emissions from these devices are defined in Part 15 of the FCC regulations [34] and Annex K of the NTIA Handbook [37]. Manufacturers of these devices and others who share these frequency bands are expected to use good design to keep both emissions and susceptibility to interference to a minimum, so that many such devices can co-exist in a noisy environment. Certain restricted bands of operation,

62

including the 4.95 GHz and GPS bands used by some LTS designs, allow only spurious signals from non-licensed devices. This does not necessarily mean that these bands are free from noise that exceeds these field strength levels. Interference may still occur from RF sources other than radiative transmitters or from changes in FCC regulations.

Table A1: Potential locator/tracking system frequency bands. Color coding in the frequency bands corresponds to the colors used in Figure 4.

LTS Band	Frequency Range (MHz)	Licensed	FCC/NTIA Rule	Other Uses of Band (47 CFR Parts - FCC)
450 MHz	450-470	Yes	FCC 47 CFR 90 Subpart K	15 – RF Devices 22 – Public Mobile 74D – Remote Pickup 74H – Low Power Auxiliary 80 – Maritime 90 – Private Land Mobile 95 – Personal Radio
700 MHz	763-775 793-805	Yes	FCC 47 CFR 90 Subpart R	15 – RF Devices 74G – LPTV and TV Translator/Booster 90R – Private Land Mobile –700 MHz Public Safety Upper Block
900 MHz	902-928	Yes No	FCC 47 CFR 90 47 CFR Parts 15, 18 (ISM)	15 – RF Devices 18 – ISM Equipment 90 – Private Land Mobile 97 – Amateur Radio
GPS L1	1563.42-1587.42	---	NTIA	87 – Aviation
2.4 GHz	2400-2483.5	No	FCC 47 CFR Parts 15, 18 (ISM)	15 – RF Devices 18 – ISM Equipment 25 – Satellite Communications 27 – Wireless Communications 74F – TV Auxiliary Broadcasting 90 – Private Land Mobile 97 – Amateur Radio 101 – Fixed Microwave
4.9 GHz	4940-4990	Yes	FCC 47 CFR 90 Subpart Y	90Y – Private Land Mobile – Public Safety use
5.1 GHz-5.8 GHz	5150-5850	No	FCC 47 CFR Parts 15, 18 (U-NII/ISM)	15 – RF Devices 25 – Satellite Communications 87 – Aviation 90 – Private Land Mobile

Table A2: Interference sources for 450 MHz **band: 450 MHz-470 MHz (Public Safety License)[1]**

Application	Power or Field Strength	Application	Power or Field Strength
On-site portable+mobile		**On-site fixed**	
Pager – Mobile station[2]	60 W ERP	Paging base station[2] (454 MHz-455 MHz / 459 MHz-460)	3500 W ERP / 150 W ERP
Mobile radio for industry/business[3]	6 W ERP	Base station for public safety or industry/business (e.g. vehicle location & monitoring) [10]	500 W ERP
Portable radio for ind./bus.[3]	2 W ERP	Fixed radio station for public safety or ind./bus. (incl. telemetry)[11]	75 W TOP
Two-way mobile radio (GMRS)[4]	5 W ERP	Low power base station for industry /business[3]	20 W ERP
Two-way mobile radio (FRS)[5]	0.5 W ERP	Signal booster for ind. / bus.[12]	5 W ERP
Wireless microphone[6]	1 W TOP	Small base station for two-way radio (GMRS)[4]	5 W ERP
Telemetry[7]	2 W TOP	Telemetry[7]	2 W TOP
Biomedical telemetry[8]	0.1 W		
Operating fire alarm[9]	12.5 mV/m at 3 m		
Intermittent transmitter (e.g. remote switch, door opener)[9]	12.5 mV/m at 3 m		
Periodic transmitter[9]	5 mV/m at 3 m		
Outside portable+mobile		**Outside fixed**	
Pager – Mobile station[2]	60 W ERP	Paging base station[2] (454 MHz-455 MHz / 459 MHz-460)	3500 W ERP / 150 W ERP
Mobile radio for industry/business (e.g. vehicle location & monitoring)[3]	6 W ERP	Rural Radiotelephone[13] (454 MHz-455 MHz / 459 MHz-460)	3500 W ERP / 150 W ERP
Portable radio for ind./bus.[3]	2 W ERP	Base station for public safety or industry/business (e.g. vehicle location & monitoring) [10]	500 W ERP
Two-way mobile radio (GMRS)[4]	5 W ERP	Fixed radio station for public safety or ind./bus. (incl. telemetry)[11]	75 W TOP
Two-way mobile radio (FRS)[5]	0.5 W ERP	Low power base station for industry /business[3]	20 W ERP
Intermittent transmitter (e.g. remote switch, door opener)[9]	12.5 mV/m at 3 m	Small base station for two-way radio (GMRS)[4]	5 W ERP
Periodic transmitter[9]	5 mV/m at 3 m		
Emergency services portable+mobile		**Emergency services base stations (outside)**	
Police or fire mobile radio[3]	6 W ERP	Base station[10]	500 W ERP
Police or fire signal booster[12]	5 W ERP	Low power base station[3]	20 W ERP
Police or fire portable radio (e.g. RF PASS, ground robot)[3]	2 W ERP	Remote pickup broadcast station for press[14]	100 W TOP
Portable remote pickup transmitter for press[14]	2.5 W TOP		
Vehicular remote pickup repeater for press[14]	2.5 W TOP		

Table A2: Interference sources for 450 MHz band (cont.)

Nearband portable+mobile		Nearband fixed	
Ind./bus. mobile radio[15] (470 MHz-512 MHz)	100 W ERP	TV channels 14-20 broadcast station[17] (470 MHz-512 MHz)	5 000 000 W ERP
Police or fire mobile radio (walkie-talkies, repeaters)[15] (470 MHz-512 MHz)	100 W ERP	DTV channels 14-20 broadcast station[18] (470 MHz-512 MHz)	1 000 000 W ERP
Biomedical telemetry[16] (470 MHz-688 MHz)	200 mV/m at 3 m	Low power TV, TV translator, or TV booster station for channels 14-20[19] (470 MHz-512 MHz)	150 000 W ERP
Wireless microphone[6] (470 MHz-608 MHz)	0.25 mW TOP	Digital low power TV, TV translator, or TV booster station for channels 14-20[19] (470 MHz-512 MHz)	15 000 W ERP
		TV STL, TV relay stations, TV translator relay station for channels 14-20[20] (470 MHz-512 MHz)	3200 W EIRP
		70 cm Amateur radio[21] (420 MHz-450 MHz)	1500 W TOP, 50 000 W ERP
		70 cm Amateur radio beacon[22] (420 MHz-450 MHz)	100 W TOP
		Public Safety or Ind./Bus. base station (e.g. vehicle location & monitoring)[15] (470 MHz-512 MHz)	1000 W
		Radiolocation Service[23] (420 MHz-435 MHz)	50 W TOP

[1] FCC Regulations 47 CFR 90 Subparts B, I, K – Public Safety Radio Pool

[2] 47 CFR 22 Subpart E (22.561, 22.565) – Public Mobile Services – Pagers (base station limits depend on antenna height and geographical location)

[3] 47 CFR 90 Subparts B, C, I, K (90.35, 90.205, 90.267) – Public Safety or Industrial / Business Radio Pool – Low power (required within 80 km of urban areas)

[4] 47 CFR 95 Subpart A (95.25, 95.29, 95.135, 95.621, 95.639) – General Mobile Radio Service (GMRS)

[5] 47 CFR 95 Subpart B (95.627, 95.639) – Family Radio Service (FRS) - Unlicensed

[6] 47 CFR 74 Subpart H (74.861) – Low Power Auxiliary Station

[7] 47 CFR 90 Subparts B, C, J (90.20, 90.35, 90.238)– Telemetry for Public Safety or Industrial / Business use

[8] NTIA Manual US209 – Frequencies for one-way, non-voice biomedical telemetry operation in medical facilities

[9] 47 CFR 15 Subpart C (15.231) – Unlicensed devices – Intermittent and periodic transmitters

[10] 47 CFR 90 Subparts B, C, I, K (90.35, 90.205) – Base station for Public Safety or Industrial / Business Radio Pool

[11] 47 CFR 90 Subparts B, C, K (90.20, 90.35, 90.238, 90.261)– Fixed station for Public Safety or Industrial / Business use (limited to 20 W TOP within 140 km of population centers)

[12] 47 CFR 90 Subparts B, C, I (90.20, 90.35, 90.219) – Signal boosters for Public Safety or Industrial / Business Radio Pool

[13] 47 CFR 22 Subpart E (22.725, 22.727) – Public Mobile Services – Rural Radiotelephone

[14] 47 CFR 74 Subpart D (74.402, 74.431, 74.461) – Remote Pickup Broadcast Stations

[15] 47 CFR 90 Subparts B, C, I, L, M (90.20, 90.35, 90.205, 90.303, 90.355)- Public Safety and Industrial/Business Radio Pool, incl. Location and Monitoring Service (LMS) – Bandwidth shared with broadcast TV (power limits depend on geographical location)

[16] 47 CFR 15 Subpart C (15.242) – Unlicensed devices – Biomedical telemetry

[17] 47 CFR 73 Subpart E (73.603, 73.614) – TV Broadcast Station

[18] 47 CFR 73 Subpart E (73.603, 73.622(f)) – DTV Broadcast Station

[19] 47 CFR 74 Subpart G (74.735) –Low Power TV, TV Translator, TV Booster Station – Analog and Digital

[20] 47 CFR 74 Subparts F, M (74.602(h)) – TV Broadcast Auxiliary

[21] 47 CFR 97 Subpart D (97.313, 97.317) – Amateur Radio Service; power limit is 50 W PEP in certain geographical areas

[22] 47 CFR 97 Subpart C (97.203) – Amateur radio beacon station

[23] 47 CFR 90 Subpart F (90.103) – Radiolocation Service – spread spectrum systems; geographical limitations

Table A3: Interference sources for 700 MHz **band: 763 MHz-775 MHz, 793 MHz-805 MHz (Public Safety License)[1,2]**

Application	Power or Field Strength	Application	Power or Field Strength
On-site portable+mobile		**On-site fixed**	
Mobile radio for public safety (e.g. security guards)[1,2]	30 W ERP	Base station for public safety– broadband[1,6]	2000 W/MHz ERP
Mobile station for 700 MHz wireless services (e.g. LTE)[3] (776 MHz-793 MHz)	30 W ERP	Base station for public safety – narrowband[2,6]	1000 W ERP
Portable radio for public safety[1,2]	3 W ERP	Fixed and base stations for 700 MHz wireless services[3] (776 MHz-793 MHz)	1000 W ERP
Portable 700 MHz wireless device[3]	3 W ERP	Control station for public safety[1,2]	30 W ERP
Mobile or portable radio for public safety – low power narrowband[2]	2 W ERP	Control station for 700 MHz wireless services[3] (776 MHz-793 MHz)	30 W ERP
Operating fire alarm[4]	12.5 mV/m at 3 m		
Intermittent transmitter (e.g. remote switch, door opener)[4]	12.5 mV/m at 3 m		
Periodic transmitter (e.g. systems check)[4]	5 mV/m at 3 m		
Wireless microphone[5]	0.05 W		
Outside portable+mobile		**Outside fixed**	
Mobile radio for public safety (e.g. security guards)[1,2]	30 W ERP	Base station for public safety – broadband[1,3]	2000 W / MHz ERP
Mobile station for 700 MHz wireless services (e.g. LTE)[3] (776 MHz-793 MHz)	30 W ERP	Base station for public safety – narrowband[2,3]	1000 W ERP
Portable radio for public safety[1,2]	3 W ERP	Fixed and base stations for 700 MHz wireless services	1000 W ERP
Portable 700 MHz wireless device[3]	3 W ERP	Control station for public safety[1,2]	30 W ERP
Mobile or portable radio for public safety – low power narrowband[2]	2 W ERP		
Emergency services portable+mobile		**Emergency services base stations (outside)**	
Police or fire mobile radio[1,2]	30 W ERP	Base station – broadband[1,4]	2000 W/MHz ERP
Police or fire portable radio[1,2]	3 W ERP	Base station – narrowband[2,4]	1000 W ERP
Police or fire mobile or portable radio – low power narrowband[2]	2 W ERP	Control transmitter[1,2]	30 W ERP
		Low power base station – narrowband[2]	2 W ERP
Nearband portable+mobile		**Nearband fixed**	
Public safety mobile radio[7] (806 MHz-817 MHz)	100 W TOP	Fixed and base stations for 700 MHz wireless services[3,4] (746 MHz-763 MHz)	2000 W / MHz ERP
Industrial/Business mobile radio[8] (809 MHz-824 MHz)	100 W TOP	Control station for 700 MHz wireless services[3] (805 MHz-806 MHz)	30 W ERP
Mobile 700 MHz wireless device[3] (746 MHz-763 MHz, 805-806 MHz)	30 W ERP		
Portable 700 MHz wireless device[3] (746 MHz-763 MHz, 805-806 MHz)	3 W ERP		

Table A3: **Interference sources for** 700 MHz **band (cont.)**

[1] FCC Regulations 47 CFR 90 Subparts B, R (90.20, 90.542) – Public Safety Radio Service – broadband (763 MHz-768 MHz, 793 MHz-798 MHz)

[2] 47 CFR 90 Subparts B, R (90.20, 90.541, 90.545) – Public Safety Radio Service – narrowband (769 MHz-775 MHz, 799 MHz-805 MHz)

[3] 47 CFR 27 Subparts C, N (27.50(b)) – 700 MHz Public/Private Partnership

[4] 47 CFR 15 Subpart C (15.231) – Unlicensed devices – Intermittent and periodic transmitters

[5] 47 CFR 15 Subpart C (15.216) – Unlicensed devices – Wireless microphones

[6] Fixed and base station limits depend on population density and antenna height

[7] 47 CFR 90 Subparts B, S (90.20, 90.635) – Private Land Mobile Radio – Public Safety Radio Pool

[8] 47 CFR 90 Subparts C, S (90.35, 90.635) – Private Land Mobile Radio – Industrial / Business Radio Pool

Table A4: Interference sources for 900 MHz **band: 902 MHz-928 MHz (LMS License or Unlicensed)**[1]

Application	Power or Field Strength	Application	Power or Field Strength
On-site portable+mobile		**On-site fixed**	
Spread spectrum unlicensed transmitters (e.g. RFID, wireless intercoms)[2]	1 W TOP, 4 W ERP	Amateur (33 cm band) station[7]	1500 W PEP, 50 000 W ERP
Zigbee Wireless PAN[2,3]	0.1 W EIRP	Spread spectrum unlicensed transmitters (e.g. RFID, wireless intercoms)[2]	1 W TOP, 4 W ERP
Field disturbance sensors[4]	500 mV/m at 3 m	Signals to measure characteristics of a material[8]	0.5 mV/m at 30 m
ISM (industrial, scientific, medical) equipment[5]	50 mW/m at 3 m	Field disturbance sensors[4]	500 mV/m at 3 m
Intermittent transmitter (e.g. remote switch, door opener)[6]	12.5 mV/m at 3 m	ISM equipment[5]	50 mW/m at 3 m
Periodic transmitter[6]	5 mV/m at 3 m		
Outside portable+mobile		**Outside fixed**	
Spread spectrum unlicensed transmitters (e.g. RFID, wireless intercoms)[2]	1 W TOP, 4 W ERP	Amateur (33 cm band) station[7]	1500 W PEP, 50 000 W ERP
Vehicle location+monitoring[9]	30 W ERP	Vehicle location + monitoring forward link[9]	300 W ERP
Field disturbance sensors[4]	500 mV/m at 3 m	Vehicle location + monitoring[9]	30 W ERP
Intermittent transmitter (e.g. remote switch, door opener)[6]	12.5 mV/m at 3 m	Spread spectrum unlicensed transmitters (e.g. RFID, wireless intercoms)[2]	1 W TOP, 4 W ERP
Periodic transmitter[6]	5 mV/m at 3 m	Field disturbance sensors[4]	500 mV/m at 3 m
Emergency services portable+mobile		**Emergency services base stations (outside)**	
Spread spectrum unlicensed transmitters (e.g. PASS device, ground robot, RFID)[2]	1 W TOP, 4 W ERP	Spread spectrum unlicensed transmitters (e.g. PASS device, ground robot, RFID)[2]	1 W TOP, 4 W ERP
ISM equipment[5]	50 mW/m at 3 m	ISM equipment[5]	50 mW/m at 3 m

Table A4: Interference sources for 900 MHz band (cont.)

Nearband portable+mobile		Nearband fixed	
Mobile station for Ind./Bus.[10] (896 MHz-901 MHz, 935 MHz-940 MHz)	100 W TOP	Fixed microwave station for Public Safety or Ind./Bus.[11] (928 MHz-932.5 MHz)	50 W EIRP
		Paging base transmitter (928 MHz-932.5)[12]	3500 W ERP
		One-way paging for Public Safety or Ind./Bus [13] (929 MHz-930 MHz)	3500 W ERP
		Fixed microwave station for Public Safety or Ind./Bus.[11] (932.5 MHz-960 MHz)	10 000 W EIRP
		Operational fixed signal boosters[14] (928 MHz-960 MHz)	5 W ERP
		Base station for Ind./Bus.[10] (935 MHz-940 MHz)	1000 W ERP
		Signals to measure characteristics of a material[8] (890 MHz-940 MHz)	0.5 mV/m at 30 m

[1] FCC Regulations 47 CFR 90 Subpart B (90.20) – Public Safety Land Mobile, 47 CFR 15 – Unlicensed, 47 CFR 18 – Industrial, Scientific, and Medical (ISM) Equipment

[2] 47 CFR 15 Subpart C (15.247) – Spread spectrum, including frequency hopping and digital modulation

[3] IEEE 802.15.4 – Low Rate Wireless Personal Area Network (ZigBee)

[4] 47 CFR 15 Subpart C (15.245) – Field disturbance sensors

[5] 47 CFR 15 Subpart C (15.249), 47 CFR 18 (18.305) – Unlicensed Industrial, Scientific, and Medical (ISM) Equipment

[6] 47 CFR 15 Subpart C (15.231) – Unlicensed devices – Intermittent and periodic transmitters

[7] 47 CFR 97 Subpart D (97.313, 97.317) – Amateur Radio Service

[8] 47 CFR 15 Subpart C (15.243) – Signals to measure characteristics of a material

[9] 47 CFR 90 Subpart M (90.205, 90.357) - Intelligent Transportation Systems (ITS) – Location and Monitoring Service (LMS)

[10] 47 CFR 90 Subpart C, S (90.35, 90.635) Industrial / Business Radio Pool – 800 MHz Cellular Service

[11] 47 CFR 90 Subparts B, C (90.20, 90.35), 47 CFR 101 (101.113, 101.147) Fixed Microwave Services for Public Safety or Industrial / Business Use – Multiple Address System

[12] 47 CFR 22 Subpart E (22.531, 22.535) Paging and Radiotelephone Service

[13] 47 CFR 90 Subparts B, C, P (90.20, 90.35, 90.494) Public Safety and Industrial / Business Radio Pool – Paging Operations

[14] 47 CFR 101 (101.151) Fixed Microwave Services – Signal boosters – Multiple Address System

Table A5: Interference sources for GPS L1 band 1563.42 MHz-1587.42 MHz[1]

Application	Power or Field Strength	Application	Power or Field Strength
On-site portable+mobile[2]		**On-site fixed**	
Spurious signals from non-licensed devices[3]	0.5 mV/m at 3 m	Spurious signals from non-licensed devices[3] GPS (DGP) re-transmitting stations[4]	0.5 mV/m at 3 m -140 dBm/24 MHz at 30 m
Outside portable+mobile[2]		**Outside fixed**	
Spurious signals from non-licensed devices[3]	0.5 mV/m at 3 m	Ground stations associated with airborne navigation aids (DGPS)[5]	Case-by-case
Emergency services portable+mobile[2]		**Emergency services base stations (outside)[2]**	
Spurious signals from non-licensed devices[3]	0.5 mV/m at 3 m	Spurious signals from non-licensed devices[3]	0.5 mV/m at 3 m
Nearband portable+mobile[2]		**Nearband fixed[2]**	
Spurious signals from non-licensed devices[3]	0.5 mV/m at 3 m	Spurious signals from non-licensed devices[3]	0.5 mV/m at 3 m

[1] FCC Regulations 47 CFR 2.106, 47 CFR 87 (87.173), NTIA Handbook M.2.1 F – Radio Spectrum Allocation, Aviation Services (Aeronautical radionavigation), Navstar Global Positioning System Bands

[2] Although the FCC permits only spurious signals from regulated sources under this category, uncontrolled sources in the environment (e.g. electrical arcing) may emit energy in this band

[3] 47 CFR Part 15 Subpart C (15.209), NTIA Manual K.3.4 – Spurious signals

[4] NTIA Handbook 8.3.28 – Use of Fixed Devices That Re-Radiate Signals Received From the Global Positioning System

[5] NTIA Handbook US343, 47 CFR 87 Subpart Q – Aviation Services – Differential GPS for aircraft navigation

Table A6: Interference sources for 2.4 GHz band: 2400 MHz-2483.5 MHz (Unlicensed)[1]

Application	Power or Field Strength	Application	Power or Field Strength
On-site portable+mobile		**On-site fixed**	
Mobile radio for industry/business or Radiolocation Service[2]	5 W	Amateur radio (13 cm band) station[10]	1500 W PEP, 50 000 W ERP
Spread spectrum unlicensed transmitters (e.g. Wi-Fi 802.11n)[3,4]	1 W TOP, 4 W ERP	Fixed microwave station[11]	30 000 W EIRP
Wi-Fi WLAN – 802.11b/g [3,4]	0.1 W EIRP	Fixed TV broadcast auxiliary station[12]	30 000 W EIRP
Bluetooth, Zigbee WPAN[3,5,6]	0.1 W EIRP	Base station for industry /business[2]	5 W
Field disturbance sensors[7]	500 mV/m at 3 m	Fixed point-to-point operations – digital modulation[14]	1 W TOP, 4 W ERP
ISM equipment[8]	50 mV/m at 3 m		
Operating fire alarm[9]	12.5 mV/m at 3 m	Field disturbance sensors[7]	500 mV/m at 3 m
Intermittent transmitter (e.g. remote switch, door opener)[9]	12.5 mV/m at 3 m	ISM equipment (e.g. microwave oven, speakerphone)[8]	50 mW/m at 3 m
Periodic transmitter[9]	5 mV/m at 3 m		
Outside portable+mobile		**Outside fixed**	
Mobile TV broadcast auxiliary station[12]	12 W TOP, 3000 W EIRP	Amateur radio (13 cm band) station[10]	1500 W PEP, 50 000 W ERP
		Fixed microwave station[11]	30 000 W EIRP
Mobile radio for industry/business or Radiolocation Service[2]	5 W	Fixed TV broadcast auxiliary station[12]	30 000 W EIRP
Radiolocation Service mobile radio[2]	1 W TOP, 4 W ERP		
Spread spectrum unlicensed transmitters (e.g. Wi-Fi 802.11n)[3,4]	1 W TOP, 4 W ERP	Base station for industry /business[2]	5 W
Wi-Fi WLAN – 802.11b/g [3,4]	0.1 W EIRP	Spread spectrum unlicensed transmitters[3]	1 W TOP, 4 W ERP
Bluetooth, Zigbee WPAN[3,5,6]	0.1 W EIRP	Fixed point-to-point operations – digital modulation[14]	1 W TOP, 4 W ERP
Field disturbance sensors[7]	500 mV/m at 3 m	Field disturbance sensors[7]	500 mV/m at 3 m
ISM equipment[8]	50 mV/m at 3 m		
Intermittent transmitter (e.g. remote switch, door opener)[9]	12.5 mV/m at 3 m		
Periodic transmitter[9]	5 mV/m at 3 m		
Emergency services portable+mobile		**Emergency services base stations (outside)**	
Police or fire mobile radio[2]	5 W	Police or fire base station[2]	5 W
ISM (industrial, scientific, medical) equipment[8]	50 mW/m at 3 m	ISM (industrial, scientific, medical) equipment[8]	50 mW/m at 3 m
Spread spectrum unlicensed transmitters (e.g ground robot, PASS device)[3]	1 W TOP, 4 W ERP	Spread spectrum unlicensed transmitters (e.g ground robot, PASS device)[3]	1 W TOP, 4 W ERP
Nearband portable+mobile		**Nearband fixed**	
ISM equipment[8] (2483.5 MHz-)	50 mV/m at 3 m	Amateur (13 cm band) station[10] (2390 MHz-2450 MHz)	1500 W PEP, 50 000 W ERP
		Fixed microwave station[11] (over 2483.5 MHz)	30 000 W EIRP

72

Table A6: Interference sources for 2.4 GHz band (cont.)

[1] FCC Regulations 47 CFR 15 – Unlicensed, 47 CFR 18 – Industrial, Scientific, and Medical (ISM) Equipment

[2] 47 CFR 90 Subparts B, C, F (90.20, 90.35, 90.103, 90.205)- Mobile radio, base station for ind/bus or radiolocation

[3] 47 CFR 15 Subpart C (15.247) – Spread spectrum, including frequency hopping and digital modulation

[4] IEEE 802.11 – Wireless Local Area Network (Wi-Fi)

[5] IEEE 802.15.1 – Wireless Personal Area Network (Bluetooth)

[6] IEEE 802.15.4 – Low Rate Wireless Personal Area Network (ZigBee)

[7] 47 CFR 15 Subpart C (15.245) – Field disturbance sensors in 2435-2465 MHz band

[8] 47 CFR 15 Subpart C (15.249), 47 CFR 18 (18.305) – Unlicensed Industrial, Scientific, and Medical (ISM) Equipment

[9] 47 CFR 15 Subpart C (15.231) – Unlicensed devices – Intermittent and periodic transmitters

[10] 47 CFR 97 Subpart D (97.313, 97.317) – Amateur Radio Service

[11] 47 CFR 101 (101.113) Fixed Microwave Services

Table A7: Interference sources for 4.9 GHz band: 4940 MHz-4990 MHz (Public Safety License)[1]

Application	Power or Field Strength	Application	Power or Field Strength
On-site portable+mobile		**On-site fixed**	
Public service mobile radio (e.g. security force) – high power broadband[1]	2.5 W TOP, 20 W ERP[2]	High power public service point-to-point and point-to-multipoint base station – broadband[1]	2.5 W TOP, 1000 W ERP[2]
Public service mobile radio – low power broadband[1]	0.125 W TOP, 1 W ERP[2]	Low power public service base station – broadband[1]	0.125 W TOP, 1 W ERP[2]
Outside portable+mobile		**Outside fixed**	
Public service mobile radio (e.g. security force) – high power broadband[1]	2.5 W TOP, 20 W ERP[2]	High power public service point-to-point and point-to-multipoint base station – broadband[1]	2.5 W TOP, 1000 W ERP[2]
Public service mobile radio – low power broadband[1]	0.125 W TOP, 1 W ERP[2]	Low power public service base station – broadband[1]	0.125 W TOP, 1 W ERP[2]
Emergency services portable+mobile		**Emergency services base stations (outside)**	
Police or fire mobile radio – high power broadband[1]	2.5 W TOP, 20 W ERP[2]	High power point-to-point and point-to-multipoint base station – broadband[1]	2.5 W TOP, 1000 W ERP[2]
Police or fire mobile radio – low power broadband (e.g. ground robot)[1]	0.125 W TOP, 1 W ERP[2]	Low power base station – broadband[1]	0.125 W TOP, 1 W ERP[2]
Nearband portable+mobile[3]		**Nearband fixed**	
Spurious signals from non-licensed sources[4]	0.5 mV/m at 3 m	Radio astronomy (4990 MHz-5000 MHz)[5]	
		Spurious signals from non-licensed sources[3]	0.5 mV/m at 3 m

[1] FCC Regulations, 47 CFR Part 90 Subpart B,Y (90.20, 90.1215) – Public Safety Land Mobile
[2] Power limits are calculated assuming a 20 MHz bandwidth
[3] Although the FCC permits only spurious signals from regulated sources under this category, uncontrolled sources in the environment (e.g. electrical arcing) may emit energy in this band
[4] 47 CFR Part 15 Subpart C (15.209), NTIA Manual K.3.4 – Spurious signals
[5] NTIA Manual US74 – Radio astronomy – This band is protected against interference

Table A8: Interference sources for 5.1 GHz-5.8 GHz bands: 5150-5250 MHz (U-NII-1), 5250-5350 MHz (U-NII-2), 5470-5725 MHz (U-NII-2 Extended), 5725-5850 MHz (U-NII-3/ISM) (Unlicensed)[1]

Application	Power or Field Strength	Application	Power or Field Strength
On-site portable+mobile		**On-site fixed**	
Radiolocation Service mobile radio[2]	Case-by-case	Amateur (5 cm band) station[8]	1500 W PEP, 50 000 W ERP
U-NII Upper (e.g. Wi-Fi 802.11a/n) – U-NII/ISM unlicensed spread spectrum (5725 MHz-5850 MHz)[3]	1 W TOP, 4 W ERP	Fixed point-to-point station (U-NII /ISM)[4]	1 W TOP, 200 W EIRP
U-NII Worldwide Low Power (e.g. Wi-Fi) (5470 MHz-5725 MHz)[4]	0.25 W TOP, 1 W ERP	Radiolocation Service base station[2]	Case-by-case
U-NII-2 Low Power (e.g Wi-Fi) (5250 MHz-5350 MHz)[4]	0.25 W TOP, 1 W ERP	Spread spectrum (e.g. Wi-Fi 802.11a/n)[3]	1 W TOP, 4 W ERP
U-NII Indoor (e.g. Wi-Fi) (5150 MHz-5250 MHz)[4]	0.05 W TOP, 0.2 W ERP	Field disturbance sensors[5]	500 mV/m at 3 m
Field disturbance sensors[5]	500 mV/m at 3 m	ISM equipment[6]	50 mW/m at 3 m
ISM equipment[6]	50 mV/m at 3 m		
Operating fire alarm[7]	12.5 mV/m at 3 m		
Intermittent transmitter (e.g. remote switch, door opener)[7]	12.5 mV/m at 3 m		
Periodic transmitter[7]	5 mV/m at 3 m		
Outside portable+mobile		**Outside fixed**	
Mobile radio for Radiolocation Service[2]	Case-by-case	Amateur (5 cm band) station[9]	1500 W PEP, 50 000 W ERP
Spread spectrum (e.g. Wi-Fi 802.11a/n)[3]	1 W TOP, 4 W ERP	Point-to-point fixed station (U-NII /ISM)[4]	1 W TOP, 200 W EIRP
Vehicle On-Board Units (e.g. for electronic toll payment)[8]	0.001 W TOP	Radiolocation Service base station[2]	Case-by-case
Field disturbance sensors[5]	500 mV/m at 3 m	Spread spectrum (e.g. Wi-Fi 802.11a/n)[3]	1 W TOP, 4 W ERP
ISM equipment[6]	50 mV/m at 3 m	Field disturbance sensors[5]	500 mV/m at 3 m
		ISM equipment[6]	50 mW/m at 3 m
Emergency services portable+mobile		**Emergency services base stations (outside)**	
Spread spectrum unlicensed transmitters (e.g. PASS device, Wi-Fi 802.11a/n, ground robot)[3]	1 W TOP, 4 W ERP	Spread spectrum unlicensed transmitters (e.g ground robot, PASS device, Wi-Fi 802.11a/n)[3]	1 W TOP, 4 W ERP
ISM equipment[6]	50 mV/m at 3 m	ISM equipment[6]	50 mW/m at 3 m
Nearband portable+mobile[10]		**Nearband fixed**	
Spurious signals from non-licensed sources[11]	0.5 mV/m at 3 m	Fixed microwave station[12] (5925 MHz-6425 MHz)	300 000 W EIRP
		Traffic flow & safety roadside units[13] (5850 MHz-5925 MHz)	2 W EIRP

Table A8: Interference sources for 5.1 GHz-5.8 GHz bands (cont.)

[1] FCC Regulations 47 CFR 15 Subpart E – Unlicensed National Information Infrastructure Devices, 47 CFR 18 – Industrial, Scientific, and Medical (ISM) Equipment

[2] 47 CFR 90 Subparts F,I (90.103, 90.205) Radiolocation Service

[3] 47 CFR 15 Subparts C, E (15.247, 15.407) – Spread spectrum, including frequency hopping and digital modulation, in 5725 MHz-5850 MHz band

[4] 47 CFR 15 Subpart E (15.407) – U-NII devices

[5] 47 CFR 15 Subpart C (15.245) – Field disturbance sensors in 5785-5815 MHz band

[6] 47 CFR 15 Subpart C (15.249), 47 CFR 18 (18.305) –Unlicensed Industrial, Scientific, and Medical (ISM) Equipment in 5725 MHz-5875 MHz band

[7] 47 CFR 15 Subpart C (15.231) – Unlicensed devices – Intermittent and periodic transmitters

[8] 47 CFR 95 Subpart L (95.639, 95.1511) – Dedicated Short-Range Communications Service (DSRCS) On-Board Units (OBU)

[9] 47 CFR 97 Subpart D (97.313, 97.317) – Amateur Radio Service

[10] Although the FCC permits only spurious signals from regulated sources under this category, uncontrolled sources in the environment (e.g. electrical arcing) may emit energy in this band

[11] 47 CFR Part 15 Subpart C (15.209), NTIA Manual K.3.4 – Spurious signals

[12] 47 CFR 101 (101.113, 101.803) Fixed Microwave Services, including TV STL stations

[13] 47 CFR 90 Subpart M (90.377) – Intelligent Transportation Systems (ITS) – Dedicated Short Range Communications Service (DSRCS) Roadside Units (RSU)

References

1. *2009 International Building Code,* International Code Council, Inc., Falls Church, VA

2. NFPA 901 *Uniform Coding for Fire Protection*, 1976 Ed., National Fire Protection Association, Quincy, MA.

3. Report of the National Fire Service Research Agenda Symposium, National Fallen Firefighters Foundation Emmitsburg, Maryland, June 1 – 3, 2005

4. http://www.wpi.edu/academics/ece/ppl/workshops.html

5. www.dhs.gov

6. http://www.justnet.org/coe_commtech/Pages/home.aspx

7. R. Fahy, P. LeBlanc. (2000, July/August). Report on 1999 firefighter fatalities. NFPA Journal 94(4), 46-61.

8. Abandoned cold storage warehouse multi-firefighter fatality fire. (1999, December). United States Fire Administration, Emmitsburg, MD.

9. Nine career firefighters die in rapid fire progression at commercial furniture showroom – South Carolina. (2009, February). NIOSH Firefighter Fatality Investigation and Prevention Program, Morgantown, WV

10. N.P. Bryner, S.P. Fuss, B.W. Klein, A.D. Putorti Jr "Technical Study of the Sofa Super Store Fire, South Carolina, June 18, 2007" NIST SP-1118 and NIST SP-1119, National Institute of Standards and Technology, Gaithersburg, MD, March 2011.

11. R. Fahy, P. LeBlanc, J. Molis. (2008, July/August). Firefighter fatalities 2007. *NFPA Journal 102(4),* 74-83.

12. Volunteer firefighter dies while lost in residential structure fire – Alabama. (2009, June). NIOSH Firefighter Fatality Investigation and Prevention Program, Morgantown, WV.

13. Career firefighter dies while conducting a search in a residential house fire – Kansas. (2011, January). NIOSH Firefighter Fatality Investigation and Prevention Program, Morgantown, WV.

14. Volunteer captain runs low on air, becomes disoriented, and dies while attempting to exit a large commercial structure – Texas. (2011, August). NIOSH Firefighter Fatality Investigation and Prevention Program, Morgantown, WV.

15. *2009 International Fire Code*, International Code Council, Inc., Falls Church, VA

16. National Fire Incident Reporting System – Complete Reference Guide, United States Fire Administration, National Data Center, Emmitsburg, MD, July 2010.

17. Uses of NFIRS - The Many Uses of the National Fire Incident Reporting System, FA 171, Federal Emergency Management Agency, United States Fire Administration, National Fire Data Center, Emmitsburg, MD, June 1997

18. M.J. Karter Jr, "Patterns of Firefighter Fireground Injuries," National Fire Protection Association, Quincy, MA, May 2009

19. Analysis Report on Firefighter Fatalities, Federal Emergency Management Agency, United States Fire Administration, National Fire Data Center, Emmitsburg, MD, August 1990

20. NFPA 901 *Uniform Coding for Fire Protection*, 1981 Ed., National Fire Protection Association, Quincy, MA.

21. Survey of Fire Departments for United States Fire Experience During 2010, National Fire Protection Association, Quincy, MA

22. M.J. Karter Jr, J.L. Molis, "U.S. Firefighter Injuries – 2009," National Fire Protection Association, Quincy, MA, October, 2010.

23. Personal communication, R. Fahy, Fire Analysis and Research Division, National Fire Protection Association, Quincy, MA, August 2011.

24. NFPA 220 *Standard on Types of Building Construction*, National Fire Protection Association, Quincy, MA.

25. NFPA 251 *Standard Methods of Tests of Fire Resistance of Building Construction and Materials*, National Fire Protection Association, Quincy, MA.

26. http://www.cdc.gov/niosh/fire

27. Firefighter Fatalities in the United States in 2003, United States Fire Administration, Federal Emergency Management Agency, Emmitsburg, MD, August 2004.

28. Firefighter Fatalities in the United States in 2008, United States Fire Administration, Federal Emergency Management Agency, Emmitsburg, MD, September 2009.

29. http://publichealth.drexel.edu/first/

30. http://www.fcc.gov/

31 J. Kent, J.R. Lawson, A.D. Putorti, "Performance of RFID Tags in Rough Duty Environments (Structural Fires and Moisture)," TN 1700, National Institute of Standards and Technology, Gaithersburg, MD, May 2011.

32 C. Bray "New York Prepares for the Fire Next Time", The Wall Street Journal Digital Network, September 2011,
http://online.wsj.com/article/SB100014240531111903392904576510280966006722.html

33. FCC Online Table of Frequency Allocations, 47 C.F.R. 2.106.
http://transition.fcc.gov/oet/spectrum/table/fcctable.pdf

34. FCC rules and regulations under Title 47 of the Code of Federal Regulations (CFR).
http://wireless.fcc.gov/index.htm?job=rules_and_regulations

35. "Understanding the FCC Regulations for Low-Power, Non-Licensed Transmitters," OET Bulletin No. 63, Federal Communications Commission, October 1993.

36. "Federal Spectrum Use Summary: 30 MHz – 3000 GHz," National Telecommunications and Information Administration, Office of Spectrum Management, 21 June 2010.

37. "Manual of Regulations and Procedures for Federal Radio Frequency Management (Redbook)," May 2011 Revision of the January 2008 Edition, National Telecommunications and Information Administration, ISBN 0-16-016464-8, May 2011.

38. A. Paul, G. Hurt, T. Sullivan, G. Patrick, R. Sole, L. Brunson, C.-W. Wang, B. Joiner, E. Drocella, "Interference Protection Criteria Phase 1 – Compilation from Existing Sources," NTIA Report 05-432, U.S. Department of Commerce, October 2005.

39. E.F. Drocella, D.S. Anderson, "Correction Factors and Measurement Procedure to Assess the Interference Impact of Linear Swept Frequency Signals on Radio Receivers," NTIA Technical Memorandum 10-464, U.S. Department of Commerce, December 2009.

40. "Impact of Emissions from Short-Range Devices on Radiocommunication Services," Report ITU-R SM.2210, Radiocommunication Sector of the International Telecommunication Union, June 2011.

41. "Impact of Power Line Telecommunication Systems on Radiocommunication Systems Operating in the VHF and UHF Bands Above 80 MHz," Report ITU-R SM.2212, Radiocommunication Sector of the International Telecommunication Union, June 2011.

42. "Questions and Answers about Biological Effects and Potential Hazards of Radio Frequency Electromagnetic Fields," OET Bulletin 56, 4th ed., FCC Office of Engineering & Technology, August 1999.

43. "Evaluating Compliance with FCC Guidelines for Human Exposure to Radiofrequency Electromagnetic Fields," OET Bulletin 65, Ed. 97-01, FCC Office of Engineering & Technology, August 1997.

44. FCC Consumer Facts – Human Exposure to RF Fields: Guidelines for Cellular and PCS Sites.
http://transition.fcc.gov/cgb/consumerfacts/rfexposure.pdf

45. "Radio-Frequency Wireless Technology in Medical Devices," Draft Guidance 1618, Food and Drug Administration, Center for Devices and Radiological Health, 3 January 2007.

46. "Biological Effects and Exposure Criteria for Radiofrequency Electromagnetic Fields," NCRP Report No. 86, National Council on Radiation Protection and Measurements (NCRP), 1986.

47. ANSI/IEEE C95.1-1992, "Standard for Safety :Levels with Respect to Human Exposure to Radio Frequency Electromagnetic Fields, 3 kHz to 300 GHz," The Institute of Electrical and Electronics Engineers, 1992.

48. K.A. Remley, G. Koepke, E. Messina, A. Jacoff, G. Hough, "Standards Development for Wireless Communications for Urban Search and Rescue Robots," International Symposium on Advanced Radio Technologies (ISART) 2007, Boulder CO, 26 February-2 March 2007.

49. DOF Defense Standardization Program: http://dsp.dla.mil/APP_UIL/displayPage.aspx?action=content&contentid=66

50. W.R. Vincent, G.F. Munsch, R.W. Adler, A.A. Parker, "The Mitigation of Radio Noise and Interference from On-Site Sources at Radio Receiving Sites," NPS-EC-10-001, Naval Postgraduate School, November 2009.

51. "Guidance on Electromagnetic Compatibility of Medical Devices in Healthcare Facilities," AAMI TIR 18:2010, Association for the Advancement of Medical Instrumentation, 8 March 2010.

52. FCC tools. http://www.fcc.gov/tools

53. Spectrum Dashboard. http://reboot.fcc.gov/reform/systems/spectrum-dashboard

54. FM Model for Windows. http://transition.fcc.gov/oet/info/software/fmmodel/

55. Radioreference.com. http://www.radioreference.com/

56. eEngineer. http://www.radioing.com/eengineer/

57. The SAFECOM Program, "Statement of Requirements: Background on Public Safety Wireless Communications," Department of Homeland Security, Vol. 1, March 10, 2004.

58. M. Worrell and A. MacFarlane, "Phoenix Fire Dept. Final Report," Phoenix Fire Department Radio System Safety Project, Oct. 8, 2004, http://www.ci.phoenix.az.us/FIRE/radioreport.pdf.

59. 9/11 Commission Report, National Commission on Terrorist Attacks Upon the United States, 2004.

60. Wireless Emergency Response Team (WERT), "Final Report for September 11, 2001 New York World Trade Center terrorist attack," Oct. 2001.

61. C.L. Holloway, G. Koepke, D. Camell, K.A. Remley, D.F. Williams, S. Schima, S. Canales, and D.T. Tamura, "Propagation and Detection of Radio Signals Before, During and After the Implosion of a Thirteen Story Apartment Building," NIST Technical Note 1540, Boulder, CO, May 2005.

62. C.L. Holloway, G. Koepke, D. Camell, K.A. Remley, D.F. Williams, S. Schima, and D.T. Tamura, "Propagation and Detection of Radio Signals Before, During and After the Implosion of a Large Sports Stadium (Veterans' Stadium in Philadelphia)," NIST Technical Note 1541, Boulder, CO, October 2005.

63. C.L. Holloway, G. Koepke, D. Camell, K.A. Remley, D.F. Williams, S. Schima, M. McKinley, and R.T. Johnk, "Propagation and Detection of Radio Signals Before, During and After the Implosion of a Large Convention Center," NIST Technical Note 1542, Boulder, CO, June 2006.

64. C.L. Holloway, W.F. Young, G. Koepke, K.A. Remley, D. Camell, Y. Becquet, "Attenuation of Radio Wave Signals Coupled Into Twelve Large Building Structures," *Natl. Inst. Stand. Technol. Note 1545*, Apr. 2008.

65. K.A. Remley, G. Koepke, C.L. Holloway, C. Grosvenor, D. Camell, J. Ladbury, D. Novotny, W.F. Young, G. Hough, M.D. McKinley, Y. Becquet, J. Korsnes, "Measurements to Support Broadband Modulated-Signal Radio Transmissions for the Public-Safety Sector," *Natl. Inst. Stand. Technol. Note 1546*, Apr. 2008.

66. W. F. Young, K. A. Remley, J. Ladbury, C. L. Holloway, C. Grosvenor, G. Koepke, D. Camell, S. Floris, W. Numan, and A. Garuti, "Measurements to support public safety communications: attenuation and variability of 750 MHz radio wave signals in four large building structures," NIST Technical Note 1552, Aug. 2009.

67. W.F. Young, C.L. Holloway, G. Koepke, D. Camell, Y. Becquet, and K.A. Remley, "Radio-Wave Propagation Into Large Building Structures—Part 1: CW Signal Attenuation and Variability," *IEEE Trans. Antennas Propagat.*, vol. 58 , no. 4, Apr. 2010, pp. 1279 – 1289.

68. K. A. Remley, G. Koepke, C. L. Holloway, C. Grosvenor, D. Camell, J. Ladbury, R. T. Johnk, and W. F. Young, "Radio Wave Propagation Into Large Building Structures; Part 2, Characterization of Multipath," *IEEE Trans. Antennas Propagat.*, vol. 58, no. 4, Apr. 2010, pp. 1290-1301.

69. W. F. Young, K. A. Remley, D. W. Matolak, Q. Zhang, C. L. Holloway, C. Grosvenor, C. Gentile, G. Koepke, and Q. Wu, "Measurements and models for the wireless channel in a ground-based urban setting in two public safety frequency bands," NIST Technical Note 1557, Jan. 2011.

70. K. A. Remley, W. F. Young, J. Healy, "Measurements of radio-propagation environments to support development of standards for RF-based electronic safety equipment," NIST Technical Note 1559, to be published, 2011.

71. L.P. Rice, "Radio transmission into buildings at 35 and 150 mc," Bell Syst. Tech. J., pp. 197-210, Jan. 1959.

72. E. Walker, "Penetration of radio signals into building in the cellular radio environment," Bell Syst. Tech. J., 62(9), Nov. 1983.

73. W.J. Tanis and G.J. Pilato, "Building penetration characteristics of 880 MHz and 1922 MHz radio waves," Proc. 43th IEEE Veh. Technol. Conf., Secaucus, NJ, 18-20 May 1993, pp. 206-209.

74. L.H. Loew, Y. Lo, M.G. Lafin, and E.E. Pol, "Building penetration measurements from low-height base stations at 912, 1920, and 5990 MHz," NTIA Report 95-325, National Telecommunications and Information Administration, Sept. 1995.

75. A. Davidson and C. Hill, "Measurement of building penetration into medium buildings at 900 and 1500 MHz," IEEE Trans. Veh. Technol., 46(1): 161-168; Feb. 1997.

76. A.F. De Toledo, A.M.D. Turkmani and J. D. Parsons, "Estimating Coverage of Radio Transmissions into and within Buildings at 900, 1800, and 2300 MHz," IEEE Personal Communications, 5(2): 40-47; Apr. 1998.

77. E.F.T Martijn and M.H.A.J. Herben, "Characterization of radio wave propagation into buildings at 1800 MHz," IEEE Antennas and Wireless Propagation Letters 21: 122-125; 2003.

78. A. Chandra, A. Kumar and P. Chandra, "Propagation of 2000 MHz Radio Signal into a Multistoreyed Building through Outdoor-Indoor Interface", Personal, Indoor and Mobile Radio Communications, 2003. PIMRC 2003. 14th IEEE Proceedings on, 3: 2983-2987 vol.3; Sept. 2003.

79. R.J.C. Bultitude, Y.L.C de Jong, J.A. Pugh, S. Salous and K. Khokhar, "Measurement and Modelling of Emergency Vehicle-to-Indoor 4.9 GHz Radio Channels and Prediction of IEEE 802.16 Performance for Public Safety Applications," Vehicular Technology Conference, 2007. VTC2007-Spring. IEEE 65th, pp. 397-401; Apr. 2007.

80. M. Karam, W. Turney, K. Baum, P. Satori, L. Malek and I. Ould-Dellahy,"Outdoor-Indoor Propagation Measurements and Link Performance in the VHF/UHF Bands", Vehicular Technology Conference, 2008. VTC 2008-Fall. IEEE 68th, pp. 1-5; Sept. 2008.

81. S.R. Saunders,K. Kelly, S.M.R. Jones, M. Dell'Anna and T.J. Harrold, "The indoor-outdoor radio environment," Electronics and Communication Engineering Journal , 126: 249-261; Dec. 2000.

82. D. Molkdar, "Review on radio propagation into and within buildings," IEE Proceeding-H, 38(1): 61-73; Feb. 1991

83. M.R. Souryal, D.R. Novotny, J.R. Guerrieri, D.G. Kuester, and K.A. Remley, "Impact of RF interference between a passive RFID system and a frequency hopping communications system in the 900 MHz ISM band," *IEEE EMC Symp.* Dig., July 2010, pp. 495-500.

84. K.A. Remley, M.R. Souryal, W.F. Young, D.G. Kuester, D.R. Novotny, and J.R. Guerrieri, "Interference tests for 900 MHz frequency-hopping public-safety wireless devices," *IEEE EMC Symp. Dig.*, Aug. 2011.

85. E. D. Mantiply, K.R. Pohl, S.W. Poppell, J.A. Murphy, "Summary of Measured Radiofrequency Electric and Magnetic Fields (10 kHz to 30 GHz) in the General and Work Environment," Bioelectromagnetics 18:563-577, 1997.

86. J. Gavan, "Main Effect of Mutual Interference in Radio Communication Systems Using Broad-Band Transmitters," IEEE Trans. Electromagn. Compat., vol. EMC-28, no. 4, pp. 211-219, June 1986.

87. T. S. Rappaport, "Wireless Communications: Principles and Practice, 2nd Edition," Prentice Hall Ptr., Upper Saddle, NJ pp 157-166, Copyright 2002.

88. J. Do, D.M. Akos, P.K. Enge, "L and S Bands Spectrum Survey in the San Francisco Bay Area," Position Location and Navigation Symposium, 2004, pp. 566-572, 26-29 April 2004.

89. D. R. Novotny, J. R. Guerrieri, and D. G. Kuester, "Potential interference issues between FCC part 15 compliant UHF ISM emitters and equipment passing standard immunity testing requirements," IEEE International Symposium on Electromagnetic Compatibility 2009 (EMC 2009), pp.161-165, 17-21 Aug. 2009.

90. C. R. Paul, Introduction to Electromagnetic Compatibility, John Wiley & Sons, Inc., Copyright 1992.

91. N. J. LaSorte, H. H. Refai, D. M. Witters Jr., S. J. Seidman, J. L. Silberberg, "Wireless Medical Device Coexistence", *Medical Electronics Design,* August, 2011
http://www.medicalelectronicsdesign.com/article/wireless-medical-device-coexistence

www.ingramcontent.com/pod-product-compliance
Lightning Source LLC
Chambersburg PA
CBHW081831170526

45167CB00007B/2787